WHEN THE LAST LION ROARS

WHEN THE LAST LION ROARS

Sara Evans

BLOOMSBURY WILDLIFE

LONDON · OXFORD · NEW YORK · NEW DELHI · SYDNEY

For my father, David Evans, whose dream was to travel to Africa and see lions in the wild. He died when he was forty-five, before he was able to travel to the continent. This book is written in his memory.

BLOOMSBURY WILDLIFE
Bloomsbury Publishing Plc
50 Bedford Square, London, WC1B 3DP, UK

BLOOMSBURY, BLOOMSBURY WILDLIFE and the Diana logo are trademarks of Bloomsbury Publishing Plc

First published in Great Britain 2018

Copyright © Sara Evans, 2018

A catalogue record for this book is available from the British Library

Library of Congress Cataloguing-in-Publication data has been applied for

ISBN: HB: 978-1-4729-1613-6; ePub: 978-1-4729-1611-2; ePDF: 978-1-4729-1612-9

2 4 6 8 10 9 7 5 3 1

Typeset in Bembo Std by Deanta Global Publishing Services, Chennai, India
Printed and bound in Great Britain by CPI Group (UK) Ltd, Croydon CR0 4YY

To find out more about our authors and books visit www.bloomsbury.com
and sign up for our newsletters

Contents

Until the lion has its own storyteller, the tale of the hunt will always glorify the hunter.

Proverb, Zimbabwe

Preface

I first saw wild lions in South Africa. I was at the Madikwe Game Reserve, in the north-west of the country. It was first light, and my first time in Africa. Bar the haunting cries of a solitary fish eagle slicing through the early morning mist, all was quiet.

The peace was soon broken by polite chatter, though, as I joined four other guests for a dawn drive into the bush. David, our guide, had just started telling us about the animals we might encounter, when some vultures circling overhead caught his eye.

'Come,' he said, 'we must head towards the vultures. They're a sign that something has been killed, and where there are dead things there may be big cats, too.'

After driving for around five minutes, we pulled up by a large acacia tree. Towards the top of the tree, straddling a branch, was a leopard. Apart from its tail flicking from side to side, slow and steady like the pendulum of a grandfather clock, the leopard was motionless. The pale fur around its mouth was stained brownish-red with blood and earth. In front of it, wedged into a fork in the branch, was a warthog, its eyes bulging, blood dripping slowly from its neck.

The leopard leaned forwards and licked blood lazily from the warthog's back. Then it stopped abruptly and looked quickly down from the tree. Not towards us in the jeep, but at two young male lions – probably brothers – that had just swaggered into view. Confident as princes, they padded slowly around the base of the tree. It wasn't just the vultures, still circling above, that had spotted the chance for a free breakfast.

The larger of the two lions raked its claws on the trunk of the tree, craning his neck in the direction of the leopard and warthog. Then he began to climb up. The leopard reacted quickly, clamping its jaws around the prize. In what seemed like a split second, the leopard descended from the tree, prey in tow.

As soon as the leopard landed, the lions moved in. Rather than head into nearby bush cover, though, the leopard moved around the acacia instead. In slapstick fashion, the lions followed the leopard around the tree. They circled madly for a while until the leopard, exhausted, dropped the warthog and retreated to the bushes.

One of the lions followed the leopard but soon returned only to find his sibling tucking into the warthog. As he approached, the lion with the warthog made a run for it, pursued swiftly by his brother. And that's the last we saw of the two marauding lions, their heist successfully completed, a culinary ambush executed perfectly, feline style.

Meanwhile, back in the jeep, we'd been library-quiet, transfixed as we watched the big-cat drama play out. We talked about what we'd just seen. I soon realised how fortunate I'd been, not just to see lions and a leopard during my first ten minutes in the bush, but to watch an entire wild story unfold before me.

Until then, the only flesh-and-blood lions I'd seen were some sleepy-looking lions at London Zoo, the dozy stars of a school trip. The others I had seen were in my imagination. There was Aslan, the magical and mysterious talking lion in *The Lion, the Witch and the Wardrobe*, and of course, Elsa, the *Born Free* lioness raised by Joy and George Adamson after being orphaned, and then returned successfully to the wilds of Kenya.

There were also the big cats on children's TV programmes from the 1960s and 70s, like *Tarzan*, *Daktari* and *Animal Magic*. Sometimes too, when I was allowed to stay up late, there were the wild lions on nature documentaries, such as *The World About Us*. Most of the lions were fictional, a few had been real, but they all felt remote and untouchable. The Madikwe lions were vivid. They were blood, tooth and claw, as real and undeniable as daylight, and I was captivated.

When I returned to England, I found myself thinking about the lions I had seen. I also thought about the leopards, the elephants, the rhinos and all the other wonderful wildlife at Madikwe. Like a holiday romance, I couldn't get South Africa out of my mind. I desperately wanted to return, to visit more African countries, to discover wildlife and landscapes I'd not seen before.

I started reading about the continent. Everything, from classics like *Out of Africa* by Karen Blixen and novels by J. M. Coetzee, to history books and wildlife narratives. But it was *Dark Star Safari* – in which the travel writer Paul Theroux describes his journey overland from Cape Town in South Africa to Cairo in Egypt – that provided inspiration on how I might be able to return to Africa.

Could I become a travel writer too, I wondered? I was then working as an editor with an educational publisher in Cambridge and in my spare time, started to work on ideas for travel pieces with wildlife at their heart.

Back then, Alexander McCall Smith's *The No. 1 Ladies' Detective Agency* books were very popular. Set in the Botswanan capital, Gaborone, the novels tell of the adventures of Mma Ramotswe, the country's first and only lady detective. I had heard of a literary tour that took visitors to the favourite haunts of Mma Ramotswe; I thought it seemed just right for a travel piece and pitched the idea to the *Sunday Telegraph*.

The editor there got back to me quite quickly. He said he liked the idea and that if I went ahead and wrote it up he would most certainly read it. Although he couldn't promise that the article would run, if it did hit the spot, the piece would be published.

I took a gamble, arranged a trip to Gaborone and booked a place on the tour. When I returned, I wrote the piece and duly sent it off. It was published a few weeks later. It wasn't about wildlife, but it was a good start, and while I was in Botswana I'd been able to visit the Linyanti Marshes in the far north of the country. There, I'd seen a soaking-wet lioness and two bedraggled cubs shaking themselves dry after swimming in nearby water, which I later learned is a rare thing to witness.

Having had one piece published, I was then able to secure further commissions, more often than not focusing on wildlife experiences. These enabled me to return to South Africa and visit other countries in the continent, including Kenya, Malawi, Namibia and Zambia.

Although during my ten years of travel around Africa, I always saw lions, the continent's big cats were actually in crisis. Yet most people seemed to be unaware that fewer than 20,000 lions remained, compared to around 100,000 in the 1990s. And that if this rate of loss continued, large parts of Africa could lose their lions for ever.

It wasn't until 2015, when a well-known lion called Cecil was killed by a trophy hunter in Zimbabwe, that people became aware of the threats faced by lions in Africa. As news of his violent death made international headlines and went viral, many were horrified to learn that the number of lions was falling so drastically that the future of the species was hanging in the balance. People expressed their anger on social media and donated significant amounts of money to lion conservation organisations. The world, it seemed, was at last sitting up and taking notice of what was happening to an animal they had assumed was invincible.

But how did it come to this? How could lions – Africa's most enduring icon, the very essence of wildness and the stalwart of fable and myth – be slipping away? We now know that lions are in trouble, but why has the species come to be in such peril? What happened to the lions that roamed Africa in their millions a couple of centuries ago? What happened to the lions that left Africa for Eurasia, eventually reaching North America? And what happened to the big cats that moved into Asia, populating vast territories from northern Greece to Pakistan and further south to India?

This book is all about these missing lions. It tells of their rise during the Ice Age and their fall in more recent times. It reflects on our relationship with them and the effect we have had on their population, from our very first encounters as early humans to the dramatic interactions of lions with pharaohs, emperors, maharajas and European royalty, to everyday folk just protecting their livestock and kin, and the trophy hunters of the modern era.

When the Last Lion Roars also tells the story of the people who are protecting our last lions, providing a safety net to help prevent further loss. These include men, women and children who share their landscapes with lions and who have found ways to live together, conflict-free.

But, perhaps most of all, this book tells the story of a world losing its wildness, unable to share what wild land remains, even if it is with an animal as charismatic and iconic as the king of beasts.

CHAPTER ONE

The Rise of the Lion Empire

It has been said that the lion's eye is not luminous; I assert it is luminous, even after death, giving forth shining light of pale green and gold, whilst in life it seems to flash forth fire.

Sir Alfred Edward Pease, *The Book of the Lion*

Chauvet Cave, Ardèche, France

It's in southern France that I come face to face with a cave lion. She's slunk low, practically on her belly, covering almost 2 metres of ground. Her rounded ears are pinned back and her front legs stretched out. Claws, long and lethal, the colour of dirty nicotine-stained nails, are extended. Mustard-seed-brown eyes stare up at me, unforgiving and focused.

She's the largest and fiercest-looking lion I've ever seen. I'm not alarmed, though, because Eve, as I have decided to call her, is a replica. From the tip of her tufted tail to the end of her whiskers, she's a wonderfully detailed reproduction of a cave lion,* displayed in an information gallery near the Chauvet Cave in Ardèche.

Evolving from earlier lion species around 300,000 years ago, cave lions like Eve roamed much of northern Eurasia. Originating from Africa, they spread into Europe, populating Spain, Germany and Britain on their way to Siberia, finally reaching North America via the Bering Strait.

---

* Historically, the Eurasian cave lion has been considered a lion subspecies, known as *Panthera leo spelaea*. It is now widely accepted that the Eurasian cave lion should be given full species status and be recognised as *Panthera spelaea*.

They were here in the south of France, too. We know this because on the walls of the Chauvet Cave there are paintings of them. Inside these towering gorges, the first modern humans, some 32,000 years ago, painted the lions they shared their landscape with.

It wasn't just lions they painted. The other fabulous megafauna that inhabited the prehistoric panorama is on the cave walls too. Woolly mammoths with their incredible tusks, giant deer with enormous antlers up to 3.5 metres wide, the woolly rhinos thought to weigh up to 2,700 kilograms and the massively-shinned cave bear are all here, depicted in such vivid detail they take my breath away.

Radiocarbon dating has revealed that the portraits of these Ice Age animals are between 32,000 and 36,000 years old, making them the earliest examples of figurative art found anywhere on earth to date. It's not just their age that's remarkable, it's their number too. Hundreds of animals (more than seventy of them lions) belonging to thirteen different species have been painted here. Small wonder the walls of the Chauvet Cave are the most decorated and most celebrated cave walls in the world.

I'm standing now in front of a spectacular mural, painted high up on the wall and almost 2 metres long, known as the Panel of the Lions. Depicting ninety-two animals including woolly mammoths and rhinos, bison and lions, its scope is staggering.

The panel tells the story of a hunt; a large group of lions chasing animals much bigger than themselves. Mammoths and rhinos are seen fleeing first, followed by an exodus of bison with lions at their rear, so close they're almost sniffing their rumps.

The lions are working together as a pride, as they would today. Their bodies are perfect prehistoric hunting machines, muscular and strong. Focused on the bison they need to bring down, the lions' shared gaze is fixed and straight ahead. Open-mouthed, the pride looks hungry, desperate for a much-needed kill.

The sense of movement in the mural is compelling. Running deftly, the lions are svelte, full of feline grace. In contrast, large lumbering bison pound the soil, the rhythm of heavy hooves matching their fearful heartbeats as they run for their lives.

That I can sense the hunger and fear of these animals, hear their faltering breath, their hearts beating hard as they run, illustrates the dexterity and creativity of the people who painted them. Perfectly observed details – like the stiffness of a prehistoric horse's mane, the closed hunch of a cave hyena's upper back and the pepper-pot sprinkling of whisker spots on a lion's muzzle – bring these beasts back from the dead.

Like prehistoric Polaroids, they add flesh and fur to the dry fossils and lifeless bones that our knowledge of Ice Age animals is largely based upon. The exquisitely executed outlines and shadings of these charcoal and ochre paintings let me see the very same animals our ancestors saw thousands of years ago, making their world and their lives feel real and immediate, less shadowy and far away.

I linger in front of another lion painting. It's one of my favourites. Almost 2 metres in length, simple and elegant, it shows two lions, a male and a female, crouching side by side. His ears are up, hers back on her head. They could be in a 'we're-ready-to-pounce-on-prey' position or they could be sniffing the ground, hoping to pick up the scent of a bison for dinner.

Or perhaps there's something more intimate, more sensual, going on? Possibly the male is lowering himself to the female's shoulder height, keen to take the lead and start a mating ritual. Whatever they're doing, there's a sense of partnership between the pair that binds them together.

Besides the beauty of the image and the artist's skill in working with the contours of the rock wall to define the lions' shape, what's really interesting is that the male has no mane. He is definitely male because he has a scrotum, drawn high up on his rump. In fact, none of the male lions in the paintings have manes, so unlike the lions that came after them, and despite the Ice Age chill, male cave lions, it appears, were without manes.

Maneless male cave lions don't just appear on the walls of the Chauvet Cave, though. Other wonderful Ice Age paintings of lions have been found in caves at Lascaux and Les Combarelles in the Dordogne and also at Les Trois Frères in Ariège, all in south-west France. Just like the big cats at Chauvet, the lions have protruding round ears and tufted tails. Most of the males are maneless, just the odd one sprouting a sketchy ruff at most.

The Chauvet Cave was discovered on 18 December 1994 by three speleologists: Jean-Marie Chauvet, Eliette Brunel Deschamps and Christian Hillaire. Walking outside around the limestone gorges here, the men detected a puff of air coming through a rock face. When they moved the rocks and stones from where the air was escaping, a small entrance was revealed.

Little did Chauvet and company know that the wisp of air coming out of the rocks like a genie from a lamp would lead them to not just a new and undiscovered cave, but a cave filled with the world's oldest and best-preserved cave art. A prehistoric time capsule filled with treasure millennia old, it was a unique legacy left behind by our ancestors some 32,000 years ago. They had stumbled upon a gift for humanity, the best early Christmas present the world has ever been given.

What the three men saw inside the cave left them speechless. Curtains of dazzlingly white calcite crystals twinkled from above, coating stalagmites and stalactites with glittering thick white dust. Strewn beneath them, like boulders of fallen cliff rock, were huge, unidentifiable bones and enormous skulls with fearsome teeth, some shrouded in calcite, sparkling like diamonds.

If the cave hunters didn't recognise the bones they saw on the ground, they certainly recognised the animals they saw painted on the surrounding walls. Charging lions, snarling wolves, cantering horses, towering mammoths and hulking bears were all there, as vivid as if they'd been painted yesterday.

Imagine the spine-tingling moment the men must have experienced when they saw and recognised the extinct wild beasts on the walls around them, when they realised they were the first people to see these fabulous creatures for tens of thousands of years.

Once the initial shock of their find had passed, the trio informed the relevant authorities about their startling discovery. Almost immediately the cave was resealed. Only masked scientists were allowed in to assess the age of the bones and paintings. The results of the carbon dating staggered everyone.

Some of the oldest paintings were found to be 36,000 years old, 17,000 years older and every bit as wonderful as the cave art (then thought to be the world's oldest) found in the Lascaux Cave. What Chauvet and his colleagues had been led to by an innocuous puff of air was the greatest Ice Age haul of cave art ever found. Prehistoric booty, precious not just for its beauty and age, but also for its incredible vulnerability and fragility.

It was decided that the cave – later named after Chauvet – should be closed to the general public. If the cave were opened, the fantastic paintings of prehistoric megafauna – as well as the tiny tentative footprint of an Ice Age human child and the small fragile bones of baby cave bears – all risked decay. The authorities had learned their lesson from what happened to the cave art at Lascaux. There, millions of visitors unintentionally introduced mould and bacteria that irrevocably pockmarked the art of our ancestors. Just by breathing, they damaged it for ever.

The only way to keep the now UNESCO-protected paintings safe while also sharing them with the world was to keep the Chauvet Cave sealed and to construct another cave, complete with duplicate prehistoric paintings and contents. Two years and €55 million later, the Caverne du Pont d'Arc, a perfect limestone doppelgänger, was born.

Although not quite as long as the Chauvet Cave, everything else inside the Caverne du Pont d'Arc is the same as the real thing. Using the same tools, materials and rock canvasses, contemporary artists have completely recreated the paintings and engravings found inside Chauvet. The scratch marks made on the walls by cave bears are here, as are the sleeping spaces they used to slumber out savage Ice Age winters. Even the temperature, light level and acoustics are just as they would have been over 30,000 years ago.

A modern masterpiece of reconstruction and technology, the Caverne du Pont d'Arc showcases the astonishing creativity of early modern humans and their appreciation of the magnificent Ice Age beasts they lived alongside. All the paintings I've seen here have been in the Caverne du Pont d'Arc. I'm glad the originals are safe in the Chauvet Cave nearby, sealed from the world with an enormous reinforced door, like the ones that guard bank vaults, protecting our heritage, priceless prehistoric heirlooms, for generations not yet born.

Outside, early summer sunlight stings my eyes. I close them and imagine what Eve would look like roaming this patch of land: a 29-kilometre stretch of deep limestone gorges, intersected by the Ardèche river, honeycombed with caves and topped off with low trees and shrubs.

Thirty-two thousand years ago, though, when glaciers over 2,500 metres thick were icing up Europe, it would have felt cooler: still sunny and dry, but with temperatures more like those that Sweden has today. The area would have been more thickly forested and around the river, on low-lying ground, would have been flat, grassy plains.

Animals would have been here too, certainly all of the ones I've seen in the Caverne du Pont d'Arc. The plains would have been covered with super-sized herbivores, ibex, giant deer and aurochs among them, grazing on succulent grasses. And where there were herbivores, there were, of course, lions.

Around 10 per cent larger than lions living in Africa today, cave lions were big and strong enough to take on plus-size herbivores like these, occasionally preying on young or injured adult woolly mammoths too. To grab and bring down such mighty prey, cave lions had huge paws equipped for the job. In 1992, a trackway used by cave lions some 35,000 to 42,000 years ago was found at Bottrop in the Emscher river valley in north-west Germany.

The trackway at Bottrop was a real find. Not only was it remarkably well preserved, but it was also – at 150 square metres – one of the longest tracks ever discovered. Described as 'true jewels of mammalian palaeoichnology' by one palaeontologist, such trackways are incredibly rare and offer valuable insights into 'the locomotion and behaviour of extinct animals'.

Prints of other Ice Age mammals, including horses, reindeers and wolves, were also captured alongside those of the cave lions, suggesting that the track was a communal path to a waterhole. When measured, the average size of the fossilised lion paw prints were recorded as between 12 and 14 centimetres, significantly larger than those of large male lions living today – with paws around 9 to 12 centimetres in size.

Although they're called cave lions,* Ice Age lions didn't actually live in caves. They were originally given this name because so many of their bones were found in caves: skulls, segments of jawbones, loose teeth, bits of spine and shards of smashed toe bones, as well as whole skeletons, have all been found littering the floors of caves throughout Eurasia.

Cave lion bones, along with prehistoric bear and wolf bones, were first discovered in 1774 at the Zoolithen Cave in Bavaria, southern Germany. They were found by Johann Esper, a local parish priest with a passion for speleology and natural history. Since then, more bones – including the remains of thirteen cave lions, mainly older males, as well as hyenas – have also been found in bone-rich Zoolithen and neighbouring caves.

---

* Historically named 'cave lions', the animals' name has recently been revised to 'steppe lion'.

Cave lion bone finds haven't been confined to Germany. Bones have been unearthed in caves and open-air sites all over Europe: in France, Poland, the Czech Republic, Austria, Croatia and Slovakia. They've even been found on the floor of the North Sea, dredged up by fishermen in 1983.

Significant finds include the skeleton of an adult male in 1985. It was found next to the remains of a woolly mammoth at an open-air site near Siegsdorf in southern Germany. When measured, it reached 1.2 metres at the shoulder and 2.1 metres in length not including its tail. Three complete lion skeletons have also been found 800 metres deep in the Urşilor Cave in the western mountains of Romania in 2009.

More recently, in the Imanai Cave in Russia's Ural Mountains, a collection of 500 cave lion bones and fragments, which could be the remains of up to 6 big cats, were discovered in August 2015 by a team of Russian scientists. Estimated to date back 60,000 years or so, this spectacular haul is believed to be unique, the world's largest cave lion bone find.

Big-cat fossils have also been found in England. As early as 1825, bones from a cave lion's jaws – believed to be 20,000 to 50,000 years old – were found at Kents Cavern in Devon in the south-west of the country by John MacEnery, a priest and archaeologist. Some 32 years later, workmen digging a canal at the Wookey Hole Caves, also in the south-west, found the remains of a prehistoric human as well as the bones of Ice Age megafauna, including lions, mammoths and rhinos, all estimated to be between 250,000 and 300,000 years old.

Up in the East Midlands, another piece of cave lion jaw – believed to be between 38,000 and 50,000 years old – was found in the Pin Hole Cave at Creswell Crags in Derbyshire by the archaeologist A. L. Armstrong sometime between 1924 and 1936. At the same site around 50 years later in 1981 another archaeologist, Rogan Jenkinson, found a lion's tooth (a canine) also estimated to be between 38,000 and 50,000 years old, when excavating the west chamber of the Robin Hood Cave.

But perhaps one of the best-known English finds is that of a cave lion's toe bone – some 125,000 years old – which was found at Trafalgar Square in London, famous for its 4 huge bronze lions that guard Nelson's Column. The toe bone was unearthed in the late 1950s, when building work was carried out at the south of the square.

It's not just cave lion bones and teeth that have been uncovered. In 2008, a local man walking by the Maly Anyuy River in Chukotka in Russia's far east came across a huge skeleton that had washed out from the perennially frozen river. Not only was this the first cave lion skeleton unearthed in Russia, but it was so well preserved it still had a clump of fur, weighing 4 grams, attached to it.

When the fur was later studied in 2016, it was compared to the fur of extant African lions. It was found to be similar in colour, though just a little lighter. The fur was very thick, thicker than the fur of today's lions, with a dense undercoat of insulating downy hair, mixed with a smaller amount of darker guard hair that acted like a barrier to protect the lion from Ice Age temperatures.

The discovery of cave lion bones has always been thrilling, but when the bodies of two perfectly preserved cave lion cubs were found in the summer of 2015, palaeontologists around the world took a deep breath. The cubs were found by contractors looking for mammoth tusks in the Abyisky district of the Sakha (Yakutia) Republic in eastern Siberia. Working by the Uyandina River, something in the riverbank caught the workers' attention. 'In one crack, we saw an ice lens with some pieces inside,' one worker told the *Siberian Times*.

What these men came across was a far cry from their usual hoard of Ice Age tusks. The two cubs were so much more than fossilised bones. Preserved by the permafrost for thousands of years, the little lions were still covered in their fur. Their ears were still there, as were their tails. Even their delicate whiskers and claws survived.

The cubs were found in such good condition that fossil felid expert Julie Meachen from Des Moines University in

the US state of Iowa told *National Geographic*: 'As far as I
know, there has never been a prehistoric cat found with this
level of preservation ... this is truly an extraordinary find.'

From the banks of the river, the cubs were taken to the
department of mammoth fauna studies at the Yakutian
Academy of Sciences in Yakutsk, the region's capital. There,
scientific tests and investigations were carried out to find out
more about the pair, now named Uyan and Dina after
the river by which they were found. Experts estimated that
the cubs were just one or two weeks old when they died. The
size of adult house cats, they each had a couple of milk teeth,
with tiny areas of blistering on their gums suggesting more
were about to come through.

The cubs' eyes were closed* and they lacked identifiable
sex characteristics, typical of newborns. Photographs of the
cubs, published by the *Siberian Times*, suggest the same. In
them, the cubs look vulnerable, just like newborn kittens and
puppies do. Their fur is a dark gold and tufty around the tips
of their tiny ears. Small patches of fur here and there are
faded, the soft yellow of a much-loved vintage teddy bear.
With their eyes shut tight and curled up in foetal positions,
the cubs look like they are sleeping.

Their legs are short, not yet long enough for walking; they
would have been barely moving around their den. Had they
lived longer, it's hard to imagine that these blind, fragile-
looking cubs who had yet to see each other would have
grown to become one of the Ice Age's fiercest predators.

Although they appear tiny, Uyan's vital statistics revealed
the Ice Age lion cubs were larger than the lion cubs born
today. Weighing in at 2.8 kilograms, Uyan was found to be
1.3 kilograms heavier than modern African lion cubs. But
while Uyan might have been heavier, the cub's tail, at 7
centimetres long, was just 23 per cent of its body length,
compared to the 60 per cent of an extant lion cub's tail.

---

* One cub, Uyan, had one eye partially open, but it's believed that
this may have happened after death.

Palaeontologists at the academy also examined the cubs on the inside. Using CT scanning, they found the digestive tract to be intact, detecting traces of liquid (most likely milk from the cubs' mother) inside Uyan's stomach.

After suckling her cubs, the mother probably left them in their den to go hunting. While she was away, the soil above the cubs, in all likelihood, collapsed on top of them, killing them instantly. Both the cubs' skeletons reveal 'fatal crushing of the skull and separation of the neck vertebrae', supporting this theory.

Falling soil could have blocked up the den's entrance, sealing the cubs off from the world. This protected the little corpses from scavengers, preserving their bodies in the permafrost for millennia, until they were unearthed 12,000 years later by men working on the land their mother once roamed.

Caves weren't the last resting places of just lions. Cave bear bones have also been found in them. For some bears, their homes were also their tombs. Many of the bones show signs of grievous harm, scars revealing details of untimely, grisly ends.

In the Zoolithen Cave, two bear skulls have been unearthed, fatally marked by killer cave lion teeth. In other caves, including Sloup Cave in the south-east of the Czech Republic, bear backbones have been found, broken and damaged, vertebrae chewed right through by lions eager to get to the warm, nutritious organs that they like to eat first.

During the brutally cold winters of the Ice Age, bears and their cubs would take to their cave beds for deep seasonal snoozing. As the boreal forest outside froze over, lions sniffed out hibernating bears, sometimes finding them tucked up inside their winter dens. In king-sized beds, lions would find 3-metre-long bear bodies, curled up and hunkered down waiting for the green of spring to arrive. Sleeping bears weren't always easy pickings for lions, though. Sometimes a bear would wake up.

Picture the scene.

A mother bear's slumber is suddenly disturbed, perhaps by the sound of something approaching, the scent of a predator filling the air. She wakes to find an enormous lion by her bed. Her heart thunders in her chest and she is filled with fear for the sleeping cubs she will protect, tooth and claw, come what may.

The frightened bear has fought off predators before, kept her cubs safe from wolves and hyenas. She can take on a cave lion too. As the lion leaps towards her, she stands on her hind legs. The huge arms that cradle her cubs tenderly become bludgeons.

The fight that ensues is bloody and messy. The air is filled with snarls, roars and growls. The lion, its brain pounded to porridge from the bear's powerful swipes to the head, falls to the ground. But the bear is not finished yet. She tramples the lion's body into the ground, breaking teeth, shattering bone, splattering the cave with marrow. Then she returns to her nest, settles her babies, closes her eyes and continues the wait for spring.

The killing of cave lions by bears explains why their bones have been found inside mountain caves. A rare few, like those discovered deep within the Urşilor Cave in Romania in 2009, are complete skeletons, resting where they fell deep inside cave bear dens.

Most are in bits, scattered about here and there. Some have been gnawed, punctured and chewed: not by the vegetarian cave bears that likely killed them, but by scavenging hyenas that also created dens inside caves.

Piles of cave lion bones have been found in hyena dens. Coming across dead lions, either inside or outside of caves, hyenas would drag their fleshy finds to their dens and lick them clean as whistles, every scrap of flesh and tendon removed. Bones, spat out, landed like jackstraws in huge messy heaps.

When the remains of lions unearthed in hyena dens have been examined from the Sloup Cave in the Moravian Karst region of the Czech Republic, at Romania's Urşilor Cave and at the Perick Caves and Zoolithen Cave in northern Germany, the bones have displayed distinct evidence of scavenging.

German palaeontologist Cajus Diedrich, who has studied the bones of cave lions found in or near cave sites in Central Europe, describes such evidence as bones being chewed on and cracked, of removed jawbones and 'brain-cases' opened up, enabling hyenas 'to feed on the brain'. Diedrich has also suggested that some lions, during times of stress, may even have been scavenged by their own kind: 'It remains a possibility that lions ... trapped in a complex cave system with several different levels might have engaged in cannibalism.'

At the Srbsko Chlum-Komín Cave in the Bohemian Karst region of the Czech Republic, an articulated and almost complete skeleton of a lioness, aged around two or three, was found in a hyena den during excavations carried out by local people between 1958 and 1972.

Studies of the skeleton have since led to speculation that the lioness may have been killed by a hyena clan occupying the den. No evidence of scavenging teeth marks were found on the bones. Instead, there is significant damage to the skull and signs of brain damage, most likely caused by fighting with another lion, quite possibly a male. A bite to the face could have left the lioness critically injured, suffering from visual and cognitive difficulties.

Perhaps the injured lioness, unable to hunt, instead tried her luck in a hyena den, looking to steal their scavenged prey or cubs. If she did, Diedrich says, the lioness would have been unlikely to survive a battle with 'hyenas defending their prey-storage' and young.

Diedrich also remarks that similar hyena behaviour occurs in Africa today, where most lion kills by hyenas are the result of the animals defending their cubs and dens. He also notes that in times of plenty, victims of antagonistic conflict are not always eaten, explaining 'the presence of complete Pleistocene [Ice Age] lion carcasses in hyena or cave bear den sites'.

Cave lion paws weren't the first lion feet to pad Eurasian soils. Around 400,000 years before cave lions graced the Ice Age, the ancestral cave lion[*] was at large, marking out territories across a broad swathe of Europe from the Iberian Peninsula in the west to Greece and the Caucasus in the east.

The oldest lion bone found in Europe to date is that of an ancestral cave lion. It washed up on the shores of the Suffolk seaside town of Pakefield near Lowestoft in the east of England. Identified as a piece of jawbone, it's dated as being 700,000 years old. In mainland Europe, the oldest lion fossil that's been found is a 600,000-year-old tooth, discovered at Isernia La Pineta in southern Italy.

Lion fossil finds this old are rare. Most are under 500,000 years old, and include bone fragments found in the Bisnik Cave in Poland and at Sierra de Atapuerca in Spain, plus a near-complete skull found in Mauer near Heidelberg in south-west Germany. Other ancestral cave lion bone finds have also been unearthed in Hungary, Greece, Italy and western Siberia.

Everything we know about the ancestral cave lions comes from fossils. There are no handy portraits on cave walls like those at Ardèche to refer to. There aren't any specimens with faded golden fur and desiccated scraps of flesh in museums to show us what they looked like. All that remain are teeth and bones, the prehistoric nuts and bolts that held ancestral cave lions together, enabling them to hunt and devour their prey.

While these mechanical bits and pieces aren't as emotive as the cave paintings we have of cave lions or the mummified cubs found in Russia, they still provide strong clues to help us decipher their lives.

---

[*] The ancestral cave lion is commonly awarded full species status and recognised as *Panthera fossilis* rather than *Panthera leo fossilis*.

For one thing, they show us that ancestral cave lions were big. Very big. Bigger than the cave lions like Eve that came after them, and around 25 per cent larger than lions in Africa today. At almost 2.5 metres long, ancestral cave lions would have been a fearful sight. Fierce and strong, as well as huge, they preyed on all kinds of enormous Ice Age megafauna, making light work of giant deer and young woolly mammoths.

Ancestral cave lions would have held their own, too, in occasional but potentially fatal bust-ups with Europe's only other top-of-the-food-chain big cat, the sabre-toothed cat. Heavy as a small horse, over 1 metre tall and with incredibly long and serrated fangs, the sabre-toothed cat used its teeth to force deadly punctures into the necks of its prey. From such severe gashes, few animals survived. Most bled to death. With their lives drained away, the super-fanged cat carried off its meal, take-away style.

In mountainous areas of Europe, ancestral cave lions slowly became smaller, evolving into cave lions as they moved into parts of northern Asia. For a time, both the ancestral cave lion and the cave lion probably lived alongside each other, the species overlapping and sharing the Ice Age landscape, until the former died out.

The ancestral cave lion is believed to have arrived in Europe via Greece from East Africa, where fossils of the world's first lion-like cats have been found. The oldest, a jawbone dated as being around 3.5 million years old, was dug up in northern Tanzania at Laetoli.

It's where British palaeoanthropologist Mary Leakey and her team of researchers discovered the footprints of three hominins (very early humans) preserved in ancient powdery volcanic ash. The way the footprints were created showed that the people who made them were walking upright. When dated, they were found to be around 3.6 million years old – the oldest ever discovered – pinpointing the time when humans first walked on two feet.

The presence of hominin footprints found in the same area as the lion-like jawbone indicates that humans were sharing the landscape with the beginnings of a big cat, a lion-in-waiting, treading time like water, until it became recognisably *leo*.

A million years or so later, something much more lion-like emerged. Its jawbone was unearthed at the Olduvai Gorge, around 45 kilometres north of Laetoli in the eastern Serengeti Plains. Examination of this jawbone revealed it to be the most definitive lion fossil yet discovered. Research also found dental similarities with jawbones of both the ancestral cave lion and the cave lion.

Descendants of this early lion likely left their East African home for Europe around 700,000 years ago. This was the start of an epic journey that eventually took lions into northern Asia, North America and the Middle East, as well as South East Asia, making lions the most widespread large land mammal, bar us, of all time.

When climatic conditions became favourable around 75,000 years ago, lions in Siberia followed their appetites in search of fresh prey a continent away. Ancestral cave lions were the first to go, leaving North Asia and crossing the Bering Land Bridge – an un-glaciated landmass of grassy steppes connecting North Asia with North America – to reach Alaska.

En route, lions shared the bridge with other migrating Ice Age traffic. Travelling in the same direction towards North America were mammoths, bison and other herbivores looking for greener, lusher grasses to chew on. Humans travelled this migratory superhighway too, eventually becoming North and South America's first settlers. Animal migrants leaving America included prehistoric horses and camels looking for new homes in Asia.

Once on the North American side, herbivores galore – reindeers, prehistoric horses and elks – grew fat on juicy Alaskan grass, perfect for feeding the great appetites of the lions that followed them there. Bison meat was also a favourite of lions. In 1979, miners looking for gold in the permafrost at Pearl Creek in Alaska came across the carcass of a bison that looked incredibly old, and which was also a startling shade of blue.

The bison, dubbed Blue Babe, was taken to the University of Alaska where a team of researchers, led by palaeontologist Dale Guthrie, carried out a series of tests revealing that the mummified beast was male and around 36,000 years old – a bison from the Ice Age. Frozen in the Alaskan ice for thousands of years, Blue Babe turned blue when chemicals in its rotting flesh reacted with the high iron content of the soil, covering the body in a coat of blue crystals.

Investigations also revealed what had killed Blue Babe. Like forensic detectives solving a murder case, Guthrie and his team found clues on Blue Babe's body that gave its killers away. Claw and tooth marks indelibly made on its rump proved that cave lions had brought down Blue Babe. When the bison was prepared for display at the university's museum, something else was discovered.

An eagle-eyed taxidermist spotted a sliver of tooth imbedded in the bison's hide. Tests on the tooth enamel revealed that it belonged to a lion, probably a scavenging one – taking an opportunistic gnaw and a chew on Blue Babe's body, losing a splinter of tooth as it ground down on the now-frozen skin and bone.

Some ancestral cave lions, despite the bountiful bison and other delicious fare, didn't stay put in Alaska. At least 340,000 years ago they headed south, deeper into America, only to become separated from lions in the north by vast impassable ice sheets that covered most of Canada, the result of a massive Ice Age freeze-up.

Cut off and isolated for thousands of years, this troop of big-cat adventurers started to change. They grew bigger, got stronger and turned fiercer, until eventually one of the largest meat-eating animals the world has ever known, the American lion,* came snarling into the world.

When the huge ice sheets that had contained them began to defrost and retreat, American lions, with the same wanderlust as their ancestral cave lion relatives, began colonising land even deeper south, reaching as far as Chiapas in southern Mexico.

As with ancestral cave lions, we don't have any cave paintings of American lions to show us what they looked like. Instead, thanks to the bubbling, viscous brown tar that has seeped up through the ground for tens of thousands of years at the La Brea Tar Pits in Los Angeles, California, we have eighty perfectly preserved American lion fossils captured in natural asphalt.

Lured in by water that lay on top of the pits, giant herbivores like the Columbian mammoth, bison and Harlan's ground sloth, entered them to drink. But instead of quenching their thirst, they found themselves stuck and sinking in dark tar, thick as treacle and as unforgiving as a lock.

As these animals struggled desperately to free themselves, huge carnivores like the American sabre-toothed cat, the dire wolf and the American lion arrived at the edge of the pools. Right in front of them were their favourite meals, there for the taking. Unable to stop themselves, they pounced on the sinking herbivores, coming to the same inglorious, sticky end themselves. In turn, they were sucked down, unable to breathe. Their final grunts, howls and growls were snuffed out as tar filled their lungs, joining the menagerie of animals trapped in the pits, bones blackening, on their way to being preserved for eternity.

There they remained entombed in the tar, undiscovered for tens of thousands of years. Every now and then bones would be found, but were generally discarded, widely

---

* Before 2014, the American lion was considered a subspecies and known as *Panthera leo atrox*. It's now commonly given full species status and referred to as *Panthera atrox*.

assumed to be the bones of animals kept at the site when it was a ranch during the 1800s.

Although the first scientific article mentioning La Brea's fossils was published in 1875, it wasn't until 1901 – when geologist William Warren Orcutt (who had collected bones from around the site) alerted the world to his finds – that the prehistoric treasure, which had been resting silently beneath La Brea's bubbling black springs, became widely known.

The first excavations took place between 1913 and 1915, when more than 1 million bones were found. Huge skulls, heavy as anchors, and backbones, long as rowing boats, were pulled from the tar. Bone by bone, Ice Age skeletons were pieced back together like jigsaws. Since then, excavations have been ongoing, the bony booty from the tar pits – some 3.5 million specimens from more than 600 species of animals and plants – now displayed at the La Brea Tar Pits Museum, built close to the site.

While La Brea has the largest collection of American lion bones in the United States, other bones have also been found in Florida, Idaho, Nebraska, Nevada and Texas. The first discovery of an American lion bone fossil was a huge (although incomplete) lower jawbone found in 1836 at Natchez in Mississippi by Joseph Leidy, a renowned fossil hunter of the time. The youngest bones yet found from an American lion are 12,877 years old, discovered over the border in Canada at Edmonton in the west of the country.

From the Natural Trap Cave in the Bighorn Mountains of northern Wyoming, parts of an American lion, around 24,000 years old, were retrieved in 1974. Like many other Ice Age animals before it, the lion had fallen into the cave's all-but-invisible entrance: an open sinkhole with a killer 25-metre-long vertical drop. Most animals that plunged into the cave died instantly. Those that survived the initial fall would have found it impossible to escape, dying soon after on the mattress of the corpses beneath them.

Located on a migration route, during the course of the last 100,000 years the cave has become full of the dead bodies of the unlucky animals that passed over it, including American cheetahs, bison and mammoths from the Ice Age. Constantly cool temperatures inside the cave have preserved many of the fallen animals almost perfectly, creating a macabre cake of fossils, layered one on top of another. The skeleton of the American lion found in the cave endured so well that signs of severe osteoarthritis were found in its leg bones and scientists were able to extract rare ancient DNA from it.

It's from such fossil finds that we know the male American lion was a third larger than lions living in Africa today. With legs longer than African lions, it was no doubt more powerful and able to run much faster, too. Up to 3 metres long, over 1 metre tall and weighing in at almost 300 kilograms, this New World lion was larger than the cave lion species in Eurasia, Alaska and the Yukon, making it one of the world's largest cats and certainly the largest species of lion to ever have existed.

Fossilised bones and teeth, preserved footprints and paintings in caves have helped us follow the journeys of Ice Age lions in Eurasia and the Americas, but unravelling the story of lions that remained in Africa – the lions that didn't stray north – is more challenging.

During the Ice Age, much of Africa was tropical. Its damp climates and soggy soils made preserving the bones of dead animals pretty much impossible, creating a lion-shaped gap in Africa's fossil record. Add this absence of fossil evidence to the large-scale eradication of lions by humans in much of their historical range, together with their wide distribution over a vast continent, and the migration history of lions in Africa becomes even harder to trace.

However, analysis of DNA from lions living in Africa today strongly suggests that – just like the ancestral cave lion – they're descended from the same early lion whose fragment of jawbone was found at the Olduvai Gorge in

Tanzania. Rather than heading towards Europe, these lions stayed in Africa, many of them leaving their evolutionary cradle in the east. Research led by Ross Barnett, an evolutionary biologist at the University of Copenhagen, published in 2014 in the journal *BMC Evolutionary Biology*, considers how Africa's changing Ice Age climate affected the migration of lions after they had become widespread throughout the continent, moving freely and occupying areas of savannah, scrub and woodland.

Sometime between 80,000 and 180,000 years ago, lions in Africa started to find their movements restricted. At that time, the climate got damper and wetter, and things became steamy and humid. Tropical rainforests grew ever more luscious and rampant, expanding from the Gulf of Guinea in the west to the Great Rift Valley in the east, turning Africa's heart green as they spread.

Rainforest roots reached deeper into the earth. Dripping branches and thickets stretched further outwards, tangling and knotting as they advanced, fast becoming an impenetrable jungle. Unable to pass through this vast living wall, lions in the south and east of the continent became isolated from populations in the north and west.

Then, after some 73,000 years of damp and humid conditions, Africa began to dry out. Rainforests shrivelled back. The Sahara, which had been retreating, marched forwards, turning savannahs and grasslands into deserts of blisteringly hot sand, cutting lions in the west off from those in the north. As lions in the west lost their connection with their northern cousins, they started moving into Central Africa, making the most of new habitats opened up by shrinking-back rainforests.

In the main, lion populations have remained separated because of natural dispersal barriers, including the Great Rift Valley that keeps lions in the east and south apart, the Sahara that kept lions in North Africa isolated from other lions in the continent, and rivers, too. Lions in Central Africa, for example, have found their range limited by the watery barriers of rivers, with the Nile cutting them off from lions in the east and the Niger isolating them from lions in the west.

Once separated, Africa's lions, over millennia, began to diverge, eventually becoming physically distinct from one another as they adapted to differing landscapes and conditions. As they did so, lion populations around Africa started to look a little different from each other. Most changes were cosmetic. Some lions grew larger and others sprouted heavier manes or developed darker coats, while other lions changed genetically, on the inside.

Historically, such morphological differences were used to divide lions into a number of subspecies − ranging from a modest two, up to a more extravagant twenty-four. Traditional scientific designations of the various subspecies of lion were based largely on geographical location, mane appearance, size and a few genetic variations, which led to eight subspecies thought to be:

- *Panthera leo leo*: the North African lion, also known as the Barbary, Moroccan and Atlas lion
- *Panthera leo senegalensis*: the West and Central African lion
- *Panthera leo nubica*: the East African lion, also known as the Masai lion
- *Panthera leo azandica*: the Northeast Congo lion, also known as the Uganda lion
- *Panthera leo bleyenberghi*: the Southwest African lion, also known as the Katanga lion
- *Panthera leo krugeri*: the Southeast African lion, also known as the Transvaal lion
- *Panthera leo melanochaita*: the Cape lion
- *Panthera leo persica*: the Asiatic lion, also known as the Asian, Indian and Persian lion.

Until early 2017, the International Union for the Conservation of Nature (IUCN) − the world's leading authority on the conservation of species – recognised two subspecies of lion as being genetically rather than morphologically distinct: *Panthera leo* (African lions) and *Panthera leo persica* (Asiatic lions).

However, after genetic studies showed that 'the split between Asian and African lion[s]' as a 'distinct subspecies' was 'untenable', lions are now split as *Panthera leo leo* (lions from India and North, Central and West Africa) and *Panthera leo melanochaita* (lions from eastern and southern Africa). Whatever their taxonomic status, though, it's agreed that after migrating from the east around 124,000 years ago, they reached all of the continent's compass points, sniffing out Africa's nooks and crannies and physically adapting to survive in their new homes.

Until its extinction in the 1950s, the north of the continent was the kingdom of the North African lion, most often referred to as the Barbary lion. Barbary males were usually considered to be among Africa's largest lions. Museum specimens suggest that males, from the top of their head to the tip of their tail, could be over 2.5 metres long. There's also a hunting record from the nineteenth century claiming one lion killed was over 3 metres long, with a serpentine tail half a metre long. Other documents record lions weighing in at between 270 and 300 kilograms. Majestic and fiercely handsome, with full manes growing around their faces like rays shining from the sun, Barbary lions are for many the archetypal lion. Wrapped in deep shaggy coats, they prowled not through open savannahs but in the forested mountains, wooded hills and coastal plains of Morocco, Algeria, Tunisia and Libya.

Among tall, aromatic Cedar trees, North African lions hunted alone or in pairs. With bellies low to the ground they stalked mountain sheep, Atlas deer and wild boars, small cones and burrs becoming tangled in their thick woolly fur as they did so.

Come the winter, when snow from northern Europe swept in over the Barbary Coast and fell on the mountains, the Barbary lion's deep, dense fur coat came into its own, keeping its stocky and well-muscled body warm in plummeting temperatures. Protected against the seasonal chill, lions in North Africa hunted throughout the winter, moving stealthily around huge evergreen trees transformed into white giants, catching snowflakes in their whiskers and frozen ear tufts.

The lions that remained in the east grew plentiful in the region, and were once found roaming in many East African countries including Burundi, Djibouti, Ethiopia, Kenya, Rwanda, Tanzania and Uganda.

Still extant in the area, albeit in much diminished numbers, lions in the east are sleeker and less hirsute than other African lions, well adapted for the scorching temperatures of the region. They're also longer-legged and have straighter backs. Both males and females stand at around 1 metre tall at the shoulder, with the heaviest males weighing in at approximately 200 kilograms.

Male lions in the east exhibit a number of mane types, although none as glamorous as that of the male lion in the north. Up in the eastern highlands, manes are thicker and heavier than those down in the lowlands, where temperatures are cooler and more humid and manes often become threadbare, sometimes non-existent.

While travelling in East Africa, I've seen golden prides of lions in Kenya's Masai Mara, sprawled out on the plains under a burning sun, just as their ancestors would have done thousands of years before them.

Masai Mara, Kenya
Driving out into the Mara, all is calm and still. We approach a lioness relaxing with a pair of cubs, one male and the other female, that look around five months old. They appear to be

settling down for a siesta. The trio loll in the afternoon sun, warming their muscles. Lying back in the grass, the lioness stretches out her legs, exposing a pale belly the colour of golden syrup, teats pink and swollen from recently suckling her babies.

The cubs look sleepy, sun-drowsy and content after a lunchtime feast, their tummies filled with chewed-up bits of warm meat. As they lick away missed titbits and flyaway gristle from their whiskers and paws, the odd flash of tiny white fangs, pointy and perfect, glint in the sun. The female cub stretches out, rubs her face up against her mother's side snugly and is soon out for the count.

Her brother rests his head on the lioness's front legs as if they are pillows. He looks up at his mother, eyelids drooping a little. Maybe she softly growls him a lion lullaby or it's the sweet warmth of her breath on his face, but he falls asleep almost instantly.

While the cubs catnap, they miss out on a treat. A lilac-breasted roller flies directly above us. This beautiful bird, the size of a crow, looks as if a rainbow has been painted onto its wings. Both males and females boast technicolour feathers of olive, turquoise, coral and tangerine. Above us, the male flies higher still. Suddenly he descends, swoops and dives, spinning like a top as he falls. The sun catches his rolling wings and he is aglow, all adazzle, a magical avian kaleidoscope.

As if picking up on our wonder for the aerial display above us, the female cub wakes up. She opens one eye and looks towards her mother, checking she's still there. Then she lifts up her chin, nose turned upwards, as if she's balancing sunbeams from the tip of it.

As she falls back into her deep whiskery dreams, the lioness watches her and takes a slow, deep breath. Relaxed, her little tawny ones are, for now, safe. Swatting the occasional fly with a practised flick of her tail, she turns towards me and stares. Her gaze feels kindly, benign, and I'm reminded of Elsa, the *Born Free* lioness in the old black-and-white photographs taken by Joy Adamson.

They share the same sated look, eyes half shut, whiskers and ears at rest, a halo of soft light around them. As the lioness keeps her gaze on me, I have what I call an 'Elsa moment'. That time when you think you feel the same thing or have a moment of recognition of something unspoken between yourself and an animal. Of course, I've no way of knowing what this beautiful Masai Mara lioness really feels, but for me it's a moment of serenity that endures.

Lions reached the heart of the continent around 73,000 years ago, occupying the savannahs, grasslands and forests of Cameroon, Chad, the Central African Republic, the Democratic Republic of Congo and what is now South Sudan.

Small to averagely sized, male lions in Central Africa weigh between 170 and 190 kilograms, and females between 119 and 175 kilograms. Muscular and lithe, they're of similar appearance to lions elsewhere in Africa.

Lions still survive in Central Africa but in tiny and dwindling numbers and are rare to find. There are only a few known museum specimens of Central African lions and none known in captivity. To get a better feel for what these lions are like, I visit the Powell-Cotton Museum at Quex Park, a stone's throw away from Birchington-on-Sea in Kent, England, which has one in its collection.

Powell-Cotton Museum, Kent, England
In the centre of Gallery Three stands a large glass case. Inside it, a male lion from the Republic of Congo grips onto a huge Cape buffalo. The lion has brought the buffalo to its knees; with its body buckled and broken, pain distorts the buffalo's face. Blood has turned the whites of its eyes red. Caught in a terrible spasm, its jaws are wide open.

The lion's right paw clasps its snout brutally, claws scratching away flesh as easily as making marks in sand. From behind the buffalo's gnarly horns, the lion plunges his teeth into the top of the buffalo's back. The lion's face is wrinkled in concentration as its mighty canines tear through the buffalo's tough hide. Blood, bright and scarlet, seeps into the lion's golden fur.

The lion and buffalo, permanently fixed in a deathly embrace, are just two of the animals that Victorian explorer Major Percy Horace Gordon Powell-Cotton sent back to the museum from his expeditions in Africa and Asia.

From around twenty-eight trips, the Major's treasures fill the museum and form a remarkable collection. Although a prolific hunter, the Major's motivation was not bloodthirsty. He believed that hunting animals to exhibit and share them with people who would otherwise never see them was a worthy thing to do. He just wanted everyone to see the wonderful creatures, then abundant, that he had been fortunate enough to behold in the wild.

The lion in Gallery Three was sent back from the Major's 1906 expedition in the east of the Republic of Congo in an area described as 'heavily populated by lions'. The reason for the lion's inclusion in the museum has a memorable story attached to it.

One morning, the Major – despite knowing that the area was thick with lions – went out on his own for a stroll. As he walked, a large male lion approached him. The Major, never without his shotgun, fired at the big cat immediately. The lion was struck, but not fatally.

It retreated, wounded, into nearby undergrowth, fallen twigs and small branches crackling underfoot. The Major followed the lion into the brushwood where he came across it stretched out on the ground. Believing the lion was too injured to respond to him, the Major leaned in close towards the lion, just a short distance between them.

The Major was wrong about the lion's condition. Despite a smashed jaw, the lion lunged forwards and brought the Major down on his back. Hearing the commotion, men from the Major's camp appeared on the scene. Despite the onslaught of spears that followed, the lion, intent on ripping out the Major's innards, would not give up his man.

Fast and scissor-sharp claws should have made light work of the Major's soft English belly. But instead, the lion, after slashing through the Major's jacket, was foiled by a copy of *Punch*, a popular satirical magazine of the time, which the Major had bundled up into his pocket.

As the lion began shredding the magazine, men with guns appeared. The lion was shot again and finally killed. Protected by the papery armour of *Punch*, the Major survived the lion's attack and was swiftly taken back to camp. There, his wife – on honeymoon with the Major – attended to her husband's seventeen wounds and nursed him back to health.

The defeated lion was sent back to England. There, at the Major's request, it was stuffed and displayed with a Cape buffalo in a grisly new battle scene. The Major's coat, ripped and shredded by the lion, is also on display at the museum.

After hearing of the Major's derring-do with the lion and how a copy of *Punch* saved his life, the magazine published the following verse:

The wounded lion with a lusty roar
Advanced to drink the gallant Major's gore;
But suffered great confusion when he felt
An unexpected Punch below the belt.
Sportsmen! herein I find a happy omen
Good for the deadly need of your abdomen.
Would you defy the foe upon his treks,
Wear Punch for armour, Punch for aes triplex.

The lion and buffalo aren't the only exhibits in Gallery Three. They share it with a whole host of other African animals, including a pair of lions from southern Africa. Presented in a panoramic display (the first ever seen in a museum), the animals are arranged like classes in a school photograph, with the tallest animals – giraffes and elephants – at the back, antelopes in the middle and little monkeys and mongooses to the fore.

There's a northern white rhino too, named after the Major, and now so rare there are only two left in the world. The animals here are so densely packed, it's as if an African Noah's Ark has recently docked, its furry passengers just disembarked.

Lions that established themselves in West Africa once enjoyed an extensive stomping ground covering around 2.4 million square kilometres of land, spanning south from Senegal in the far west up to the Democratic Republic of Congo in the centre of the continent. They also established bases in the countries in between, including Burkina Faso, Côte d'Ivoire, Ghana, Benin, Nigeria, Niger and parts of Chad and the Central African Republic. Today, their range is greatly reduced and they are dwindling in number, just like their cousins in Central Africa.

They're also similar in size to Central African lions and are generally smaller than lions living in the east and south of the continent. Many males have weak, barely-there manes, and some, like the cave lions of Eurasia and North America, have no mane to shake at all.

Finding conditions favourable in the south, lions that reached the tip of the continent soon established themselves in the grass-, shrub- and woodlands of Angola, Namibia, Zambia, Zimbabwe, South Africa, Botswana and Mozambique, among other southern African countries.

Generally, southern African lions are larger than lions elsewhere in the continent. Historically, some have been recorded as reaching remarkable sizes. Hunting records from the 1930s claim that a lion shot in the former Transvaal province of South Africa weighed over 300 kilograms, while a lion killed in southern Angola in 1973 was recorded as measuring just shy of an incredible 4 metres, longer than the American lion of the Ice Age!

Descendants of these early southern settlers still roam the south, but not in the numbers they once did. In South Africa, a distinct population of lions has become completely extinct. This lion, popularly known as the Cape lion, reached Africa's most southerly point, establishing a stronghold around the Province of Cape of Good Hope and Cape Town, the region's capital.

With a body that could reach over 3 metres in length and a dark, full-volume mane that grew down to its belly, the male Cape lion was the largest, heaviest and perhaps most majestic-looking lion south of the Sahara.

Isolated from other South African lion populations by the 1,046-kilometre-long Drakensberg escarpment and much larger than other southern African lions, the Cape lion was historically considered genetically distinct. However, testing of DNA from the bones of a museum specimen in 2006 proved this not to be the case.

Rather, the lion of the Cape grew darker, bigger and stronger, and as a result more spectacular, as it adapted to hunt and survive in its southerly environment.

Lions in Asia once also had a vast, wild kingdom – one that stretched from northern Greece into Albania, Bulgaria, Macedonia and Serbia, across to West Asia via Turkey, Saudi Arabia, Iran, Iraq, Syria and Oman, alongside outposts in Afghanistan, Bangladesh, India and Pakistan in South Asia.

Around 21,000 years ago, an exodus of lions left their homes in North Africa. Affected by periods of severe drought, they moved into Asia via northern Greece, looking for lands less dry where more prey could be found. Five thousand years later another group of African lions also left the north, this time settling in the Middle East.

Their adventures didn't stop there, though. Research from 2011 examining the genetic diversity and evolutionary history of lions in West and Central Africa came to the conclusion that descendants of Asiatic lions, living near India, returned to their roots in Africa. Finding the once arid areas in the west and central regions now habitable, they recolonised the region.

This return voyage helps to explain the close genetic relationship that DNA testing has shown to exist between lions in West and Central Africa and the 500 or so lions that remain in India today. It also explains why lions in India and lions from West and Central Africa share physical similarities too.

Like lions from West and Central Africa, Asiatic lions are also smaller than other African lions in the east and south of the continent. Males hit the scales at between 160 and 190 kilograms, while females weigh in at between 110 and 120 kilograms. Height-wise, they stand at just over 1 metre tall.

Unlike most African lions, Asiatic lions display greater variety in the colour of their coats. Instead of a uniform shade of mellow gold, the coats of these lions come in a hotchpotch of shades, from a dark sandy yellow through to an autumnal russet brown, with occasional black speckles. Others have ashy-grey fur coats that take on a silvery sheen when the light catches them.

Males have mainly dark brown and black-looking manes that tend to be scrawny and sit flat on the head. Like unkempt sideburns, scraps of scanty mane also grow around the males' cheeks and throat. Both males and females have a longitudinal fold of skin, rarely seen in African lions, that runs down the centre of their stomachs.

For millennia, it seemed that lions were almost everywhere. From the tip of southern Africa through northern Eurasia to North America and South Asia and the Middle East, lions had the most extensive distribution of any un-winged mammal, bar us, on earth. Fantastic opportunists, lions found and adapted to new environments with unprecedented success.

Able to live through cruel Ice Age winters, survive at high altitudes and adapt to most habitats, except the most inferno-like deserts or congested rainforests, lions at one time seemed invincible, as if they could hold the world in their paws.

Africa roared with lions. Tens of thousands of lions lived in the far north and tens of thousands more roamed the south. From the east across to the west, lions filled the continent, leaving only its most inhospitable, most life-defying regions unconquered.

But where lions once wandered, they are no longer found. However rampant they once were, most of the world no longer hears them roar. Their empires in Europe, Asia and the Middle East are gone. In Africa, their kingdoms are greatly diminished and many lions live on the edge of extinction, just a whisker away from leaving us for ever.

CHAPTER TWO
The Fall of the Lion Empire

The lion is, however, rarely heard – much more seldom seen.

John Hanning Speke, *The Discovery of the
Source of the Nile*

While they have been on earth, lions have passed over much of it. They have scaled mountains and crossed land bridges. They have outwitted droughts millennia long and waited out the wilting of vast rainforests continents wide. In a historic range that once spanned from South Africa to Morocco, Libya to Greece, Italy to Siberia, Alaska to Mexico, Armenia to Iraq and Iran to India, caravans of lions once turned the world golden.

As lions travelled the world they adapted to new habitats where they survived and thrived. Burning brightly, they remained top-of-the-food-chain predators wherever they roamed. At one time, there may have been so many lions scattered across the globe that counting them would have been impossible, simply unimaginable – no one could live long enough to count all the world's lions.

Since those heady days, both the range and number of lions has declined so drastically that conservationists and scientists only need to count lions south of the Sahara in Africa and in one patch of forest in India to assess how many we have left.

Long gone from Eurasia and North America, only fossils and cave paintings, millennia old, can tell us what the lions that roared their way through the Ice Age looked like. And in North Africa, the Middle East and South West Asia, just a handful of people are able to tell the tale of having wild lions in their midst.

In the 1880s, around 1.2 million lions are believed to have been at large in Africa. Fast forward to the 1950s and some scientists suggest the number had fallen to 500,000. By the 1990s, there were just 100,000. Today fewer than 20,000 are estimated to remain. Across the Indian Ocean, a tiny enclave of just 500 or so lions survive at India's Gir National Park, the last members of a group that once existed in their hundreds of thousands in a range spanning as many miles.

Every single day the numbers of living lions fall. One by one, pride by pride, they are disappearing from view. The lion's share is no longer lion-sized. In a very short space of time – some believe just 75 years – around 90 per cent of the world's lions just aren't here any more.

The first lions to exit the world stage were the biggest big cats: the ancestral cave lion, the cave lion, and the biggest big cat of them all, the American lion. The ancestral cave lions and the cave lions that stalked the landscapes of Ice Age Europe and northern Asia went first, leaving the planet at around the same time, approximately 14,000 to 14,500 years ago.

Map 1: Historic range of Ice Age lions

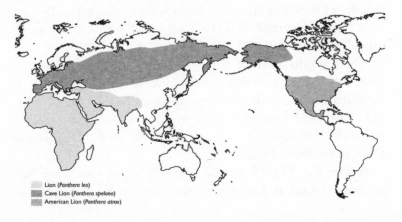

Lion (*Panthera leo*)
Cave Lion (*Panthera spelaea*)
American Lion (*Panthera atrox*)

In 2006, a six-year project called the Late Quaternary Megafaunal Extinctions (LQME) project was initiated by the Natural History Museum in London, England. It was led by Adrian Lister, a senior palaeobiologist based at the museum. The project's main objective was to determine why so much megafauna – including cave lions and bears, woolly mammoths and rhinos, giant deer and spotted hyenas – went extinct in Eurasia and North America during the latter stages of the Ice Age.

Another of the project's aims was to establish likely extinction (termination) dates for these large mammals. To do this, the bones and teeth unearthed at fossil sites across the former ranges of each species were sent to the Oxford Radiocarbon Accelerator Unit at the University of Oxford to be dated using the latest technology. Once accurate dates were determined, scientists involved in the project plotted them into maps. Using the dates from the youngest specimens, they were then able to estimate when the animals became extinct and what might have caused their demise.

Towards the end of the project, Lister and fellow project member Tony Stuart from Durham University published the paper 'Extinction Chronology of the Cave Lion', which detailed their results regarding the extinction dates for both Eurasian and North American Ice Age lions.

In total, 111 dates were established for fossilised cave lion bones and teeth. These dates were then plotted chronologically, enabling Lister and Stuart to demonstrate 'unequivocally that cave lion[s] survived into the Lateglacial across much of its range', 14,000 to 14,500 years ago.

Their results showed that in Europe, the youngest cave lion fossil was a tooth discovered in a quarry site at Sigmaringen in southern Germany, found to be 14,141 years old. The next youngest remains were bones some 14,378 years old, belonging to a lion unearthed at the Le Closeau Cave in northern France.

Before the LQME project started, a piece of cave lion jawbone found at Lathum in the eastern Netherlands had been estimated as being 10,670 years old and was considered the youngest cave lion fossil ever found. However, when tested again as part of the LQME project, it was dated as being 45,553 years old, around 30,000 years older than the lion bones uncovered at the Le Closeau Cave in France.

Test results from lions found further north showed that Eurasia's last cave lion died 14,640 years ago at the Lena Delta in the Sakha Republic, in the far north of eastern Siberia. The second youngest remains, discovered by the Arga-Yurekh River at Magadan, also in Siberia's far east, were dated as being just fifteen years older.

These results show that from the tip of Siberia down into Western Europe, cave lions 'disappeared more or less synchronously' with the 3 dates for Western Europe's youngest specimens falling between 14,100 and 14,900 years ago, and the 2 youngest dates for the 2 finds in north-eastern Siberia falling between 14,600 years and 14,800 years ago.

Radiocarbon testing of North American cave lion bones showed that the 2 youngest specimens were 13,300 and 13,800 years old, indicating that in Alaska and Yukon, ancestral cave lions and cave lions hung on a little longer than their Eurasian counterparts – their roars were heard in North America for another thousand years or so.

This is actually thousands of years earlier than the widely reported terminal date of 10,370 years attributed to a lion bone excavated at a placer gold mine site in Lost Chicken Creek in east-central Alaska. This date is now considered invalid, not just because of dating errors but because the bone actually came from a bison, not a lion. After cave lions in North America became extinct like their Eurasian cousins, the only lions left snarling on the continent were the enormous American lions.

Another study led by evolutionary biologist Ross Barnett dated the youngest American lion remains, unearthed at a pit in Edmonton in northern Canada, as being 12,877 years old, 550 years younger than an American lion toe bone retrieved from the Jaguar Cave in Idaho in north-west America, previously thought to be the youngest-known specimen.

Based on this date, it appears that American lions survived four centuries longer than the cave lions with which they once shared North America.

Fang-heavy and loaded with muscle, these extraordinary Ice Age cats, with their plus-size jaws and killer-heavy paws, were the last lions to walk on northern Eurasian and North American soils. No other wild lions have walked on the lands of their former kingdoms since.

Just as Ice Age lions vanished from Eurasia and North America millennia back, lions in much of Africa are vanishing today. The situation is so grave that lions in Africa have been listed on the IUCN's Red List of Threatened Species since 1996. In West Africa, lions have been listed as critically endangered since 2015.

Working closely with experts and conservation specialists, the IUCN regularly evaluates census data to assess how lions

Map 2: Historic and current range of lions in Africa

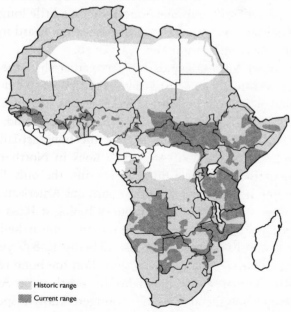

    Historic range
    Current range

are faring in the African continent as a whole, as well as in individual countries. One of the most recent sources of census evidence used by the IUCN to evaluate the current crisis faced by African lions was a 2012 study called 'The Size of Savannah Africa: A Lion's (*Panthera leo*) View'.

Described by *National Geographic* as 'the most comprehensive assessment of lion numbers to date', the work was led by Jason Riggio from the School of the Environment at Duke University in North Carolina in the United States, and co-authored by twelve scientists from America, Africa and the United Kingdom.

Keen to collect as much accurate information about contemporary lion populations as possible, the study took a new approach to gathering and analysing data. In the first instance, the focus was on establishing how much land in Africa was actually available to support lions.

To determine this, a team of graduate students from Duke University, managed by Riggio, studied the latest high-resolution Google Earth satellite images of savannah areas in Africa. Google Earth images were specifically used because previously used low-resolution satellite images were often unable to identify small areas of fields and human habitation that threaten lions' survival.

In addition to assessing how much viable lion habitat remained intact, existing continent-wide lion census data was also re-evaluated. This part of the research was managed by Stuart Pimm, professor of conservation ecology at Duke University. He brought together the study's co-authors, who all had experience preparing and contributing the data they were tasked with evaluating.

The co-authors also expanded on earlier research with additional lion counts from more than forty country-specific reports, as well as data sourced from their own studies. Merging this data with the results of the Google Earth image analysis created an 'updated geographical framework', enabling the team to estimate the number of remaining lions and the amount of suitable land available for them to live on.

The results of the study were shocking. Their estimates suggested that only 32,000 lions remained in Africa, perhaps 35,000 at the very most, down from around 100,000 lions just 50 years ago, with just 9 countries – the Central African Republic, Kenya, Tanzania, Mozambique, Zambia, Zimbabwe, South Africa, Botswana and Angola (possibly) – containing at least 1,000 lions. Tanzania alone contained 40 per cent of Africa's lions.

Across the whole of the continent, the study showed that only sixty-seven areas of land providing viable habitat for lions to live on remained. Of these areas, just 10 lion strongholds (areas with populations, usually of 500 or more, deemed most likely to survive) were identified. Every stronghold was located in southern and eastern Africa: in Botswana, Kenya, Mozambique, Tanzania, South Africa, Zambia and Zimbabwe, many of them in national parks.

No strongholds were identified in West or Central Africa. Of Africa's 32,000 surviving lions, only 24,000 were identified as belonging to viable populations. Pimm explained the severity of the situation to *National Geographic*, noting that more than 5,000 lions lived in small and isolated populations and that lions in 'West and Central Africa are in the gravest danger of extinction. More than half of the populations vital to lion conservation in these regions (as noted by the IUCN) have been extirpated in the past five years, with several countries losing their lions entirely.'

While speaking with *National Geographic*, Pimm also spoke about the difficulties of accurately estimating lion populations. 'Lions are difficult to count even though they are social and sleep most of the day,' explained Pimm. Capturing them clearly on camera can be challenging, as can finding lions in the first place, especially when they are shy and able to cover long distances.

A number of methods are used to count lions. The most common are track and call-up surveys. Track surveys involve recording lion footprints, while call-up surveys rely on lions responding to broadcast sounds and being counted as they appear. They're also counted by researchers looking for lions either on foot or in vehicles, and occasionally in helicopters. In the absence of actual sightings, lion spoor and pugmarks are also used as indicators of lion numbers.

A more accurate but expensive method involves photographing lions with high-resolution cameras. Sometimes bait is put out to lure lions into the open. Then, as each animal tucks in, it is photographed from both sides to ensure that each lion snapped and counted is unique. Remote cameras are also used.

None of these methods are ideal. Pimm describes them as 'imperfect techniques' and other conservationists have similarly described them as 'too inaccurate and too imprecise'. However, in 2017 a study called 'Toward Accurate and Precise Estimates of Lion Density' was published in the journal *Conservation Biology*, the results of which suggested that a more effective method could be used to gather accurate lion census data.

Led by Nic Elliot, project director of the Kenya Wildlife Trust's Mara Lion Project and postdoctoral researcher with Oxford University's Wildlife Conservation Research Unit (WildCRU), lions were counted at the Masai Mara National Reserve and adjacent conservancies in south-western Kenya. Over 90 days, Elliot's team covered around 8,529 kilometres, spotting and photographing 203 individual lions, close up and in high-resolution format. Each lion's unique whisker spots, acting as a form of identification, were logged. This data was then analysed using a tailor-made computer model that used a 'spatially explicit capture-recapture method', which takes into account that not all of the lions in the surveyed area may have been seen and photographed.

Elaborating on this method, co-author Arjun Gopalaswamy from the Indian Statistical Institute and the Department of Zoology at Oxford University, stressed that 'a survey might reveal there are 200 identified lions, but it will tell you nothing about how many were missed and where. Our method crucially corrects for this problem ... by estimating density at a very fine scale so that we can produce a map to show which areas have high or low density.'

As well as calculating densities, the study's capture-recapture method works with other information, including telemetry, genetic data, sex ratios and sex-specific home range sizes, providing further information about the long-term viability of surveyed populations.

Compared with track and call-up approaches, described by Elliot and Gopalaswamy as 'providing misleading estimates of population trends', the researchers concluded that their survey method was 'the most robust yet described'. As a result of their research, the pair have called for their counting method to be used across the continent in all future lion surveys to create a common accurate framework in the future.

Regardless of the approach used, surveys work best when individuals in lion populations are known to the researchers and scientists counting them. This makes identifying lions easier and helps to avoid the animals being counted twice. However, studying lions and learning to identify them accurately takes time and requires substantial resources. Only a few lion populations, including those at the Masai Mara National Reserve in Kenya and Liuwa Plain National Park in Zambia, are known at this level.

In some African countries like Angola, the Central African Republic, Ethiopia, Somalia and South Sudan, there are no reliable numbers for lions at all. In countries like these, made fragile by conflict and poverty, there are simply no funds available for experienced conservation staff, high-resolution cameras, smart mapping devices and reliable and robust vehicles to survive the gruelling long distances often involved in counting lions.

Even in countries like Zambia and Tanzania that have well-established traditions of conservation research, large areas of lion country have not been surveyed. And in some countries where data is available, it's not always accurate or useable. And, of course, as soon as numbers are published, they are out of date and inaccurate. As African lion expert

Sarel van der Merwe told *The Guardian* when interviewed about declining lion numbers, 'Counting lions to the very last individual is humanly impossible. They are difficult to count, and the finances involved simply do not allow such venture.'

The researchers behind 'The Size of Savannah Africa: A Lion's View' also note how the use of different census methods and the surveying of different geographical areas by scientists makes comparing estimates difficult. For example, Table 1 shows the different numbers estimated by Philippe Chardonnet in 2002, Hans Bauer and Sarel van der Merwe in 2004, the IUCN in 2006 and Riggio et al. in 2012, as cited in their own survey 'The Size of Savannah Africa: A Lion's View'.

However, as Riggio et al. point out, even though 'the estimates are broadly similar, there is much evidence of [lion] population decline and little to support any population increases'. Because of these difficulties, no absolute number for African lions exists today or in the past. Even the much-quoted estimate of 100,000 lions made around 50 years ago was estimated 'using rough calculations of the size of remaining habitat and density', and the IUCN's Cat Specialist Group's estimate of between 30,000 to 100,000 lions in Africa made in the 1990s, is acknowledged to be an educated guesstimate.

Table 1: Lion numbers by region and source

Region	Chardonnet (2002)	Bauer and van der Merwe (2004)	IUCN (2006a, b)	Riggio et al. (2012)
West	1,213	701	1,640	525
Central	2,765	860	2,410	2,267
East	20,485	11,167	17,290	18,308
South	13,482	9,415	11,820	11,160
Total	37, 945	22,143	33,160	32,260

The commonly cited figure for the remaining lions in Africa and the one used by the IUCN is 'fewer than 20,000'. Although precise numbers vary, that a shocking decline is happening is not disputed. Everyone agrees that lions are vanishing from the continent at an unprecedented rate.

In October 2015, yet more bad news about the state of Africa's lions was published. Eight well-established lion scientists and researchers from the universities of Oxford and Minnesota, the Grimsö Wildlife Research Station, the IUCN Species Survival Commission Cat Specialist Group, and Panthera, the global wild cat conservation organisation, analysed data regarding forty-seven lion populations across Africa and found that most were falling rapidly.

To establish rates of decline in lion populations around the continent, the team compiled data from all credible lion surveys since 1990 and then used a scientific probability model to 'calculate growth rate per population, and estimate broader trends per geographic region' for the next twenty years.

The study, called 'Lion (*Panthera leo*) Populations Are Declining Rapidly across Africa, Except in Intensively Managed Areas', revealed that lion numbers around most of the continent were dropping drastically and that without intervention would decline by a further 50 per cent over the next two decades.

In all of Africa, only four countries had lion populations that weren't in free fall. These countries – Botswana, Namibia, South Africa and Zimbabwe – were all in the south of the continent, where most lions live in protected areas.

Everywhere else, especially in West and Central Africa, the outlook for the continent's lions looks grim. The study's population model indicated that lions in the east have a 37 per cent chance of their numbers falling by half during the next twenty years. And in West and Central Africa, lions – including those in national parks at Waza in Cameroon, Niokolo-Koba in Senegal, and Kainji and

Yankari in Nigeria – have a 67 per cent chance of their numbers halving over the same time period.

The study also found that smaller lion populations, particularly those living outside of protected areas, were especially vulnerable and will likely disappear altogether. It also suggested that lions at Comoé National Park in Côte d'Ivoire and Mole National Park in Ghana are probably already extinct.

It was the findings from this report that prompted the IUCN to regionally assess lion numbers to reflect the population disparities in different areas of the continent. As a result, lions in West Africa were classified as critically endangered in 2015. The listing of lions in East and Central Africa as regionally endangered and those in the south as being of least concern is currently in preparation.

After the study's publication, Luke Hunter, co-author and Panthera president, commented on the implications of the research's stark findings in a Panthera press release: 'We cannot let progress in southern Africa lead us into complacency. Many lion populations are either gone or expected to disappear within the next few decades. The lion plays a pivotal role as the continent's top carnivore, and the free fall of Africa's lion populations we are seeing today could inexorably change the landscape of Africa's ecosystems.'

In the north of the continent there are no lions left to count. Once upon a time, though, North African lions, most often referred to as Barbary lions, lived in the region in their thousands. They hunted in the forests of the Atlas Mountains, stalking prey on the edges of Egypt's desert areas and roaring at the waves crashing in from the Mediterranean Sea and the Atlantic Ocean on the coastal plains of Morocco, Algeria and Tunisia.

There are no historic survey results to tell us exactly how many lions once lived in North Africa. They became extinct in the wild long before lion populations around the continent began to be counted. Instead of census data, there are records of sightings, dating back to the 1830s, that help pinpoint when the last Barbary lions walked the earth.

In the north of Morocco, Barbary lions were commonly seen around the Mediterranean and Atlantic coastal areas between the sixteenth and eighteenth centuries. Further inland, they were spotted by Berbers riding among the wild pear trees of the Marmora forest and oak trees of the Rif mountains up until the 1880s.

But as the 1800s drew to a close, sightings of lions in Morocco became fewer as they moved away from peopled places, taking cover in the Atlas Mountains and around the dusty edges of the Sahara. In Tunisia, lions are believed to have been lost by 1891, although stories of sightings in the Kroumir Mountains and Fériana in the north-west continued into the 1900s. In Algeria, Barbary lions and their cubs were being taken from the wild well into the 1890s. The country's last lion is generally thought to have died in 1893.

In 2013, a research paper called 'Examining the Extinction of the Barbary Lion and its Implications for Felid Conservation' was published by *PLOS ONE*, a peer-reviewed scientific journal. Led by Simon Black from the Durrell Institute of Conservation and Ecology at the University of Kent, England, one of the paper's main objectives was to determine a more accurate termination date for the lions of North Africa.

Prior to Black's research, the year generally quoted for the extinction of the Barbary lion had been 1942, when the last lion was shot and killed in Morocco. To assess the accuracy of this date, Black and his three co-authors gathered, read through and logged first- and second-hand accounts of lion sightings in North Africa, beginning in 1832. They also interviewed people who recalled seeing lions during their childhood or who were able to relay details of encounters with lions passed on to them by family members.

The accounts were logged by date, location and witness name (if known), and make for interesting reading. Sightings by hunters are included, as well as details of their lion kills. For example, a hunter called F. Chassaing is cited as seeing lions at the Aurès Mountains in north-eastern Algeria, where he shot and killed thirty lions during his lifetime. And another hunter, an Ahmed Ben Amor, is documented as seeing lions in Tunisia in 1856 and 1857, with a note stating that he killed thirty-eight lions during his hunting career.

Other valuable accounts include a lion seen attacking a sleeping man near Bou Saâda in north-central Algeria in 1840, and a lion spotted fleeing a burning Algerian forest in 1866 at Djebel Tangout. Later on, in 1935, in the north of Algeria, at Djebel es Somm, a man called Hamami Bachir is noted as seeing a male lion attacking a camel.

From France, there are colonial witnesses too. In Algeria, the botanist the Vicomte de Noé is recorded as seeing lion tracks in the Aurès Mountains in the mid-1880s, and the Marquis of Segonzac, a French army officer and explorer, is cited as seeing lions at Budaa woods at Azrou in the Berber region of Morocco in 1901.

As well as interviews and published accounts, the collection also includes the photograph in the colour section of this book. It was taken from the air by a French photographer called Marcelin Flandrin as he passed over a lone male Barbary lion on the Casablanca–Dakar air route in 1925, and is the last visual record of a wild lion in North Africa.

After reviewing their compiled evidence, Black and his team were able to determine that the last sighting of a Barbary lion in North Africa was later than the previously recorded date of 1942.

Black and his team logged a reliable sighting made in 1956, some fourteen years after North Africa's last lion was believed to have fallen. The lion was seen by several people travelling by bus through the forest, just north of the town of Sétif in the east of Algeria.

In his academic blog about Barbary lions, Black also notes that this forest 'was destroyed in 1958 during the French–Algerian War, so it is possible the last lions disappeared at that time' once their habitat was gone. As well as establishing a later extinction date for North African lions, the sightings documented in the research also indicate how people affected lions' movements as they left populated areas. For example, in Morocco lions travelled 'southwards away from coastal regions through the Rif, Middle and High Atlas Mountains and the Saharan fringes', while in Algeria they headed east 'into the Tell Atlas and the Aurès Mountains bordering Tunisia'. In both of the regions, the small populations were able to survive for several generations.

Since 1956, no wild lions have been seen in the north. No signs of hunting in the forests. No pugmarks in the sand. No roars heard by the sea. North Africa is without lions, a lion-free zone, much the same as Eurasia and the Americas. Although Barbary lions may be gone from the wild, an estimated eighty or so of their possible descendants survive. Likely to have, at some time, been mixed with lions from Central Africa, they live in captivity at various zoos around the world, including the Port Lympne Reserve in south-east England.

It's here that I see two lions, a male and a female called Milo and Ruti. Both are believed to have some Barbary heritage, since they're each identified in the Moroccan Royal Lion Stud Book as descendants of a collection of North African lions with wild origins once owned by the King of Morocco.

As part of an official European captive breeding programme to help preserve a possible Barbary lion gene, Ruti was brought to the south of England from a Spanish zoo in Madrid to breed with Milo in 2011.

Port Lympne Reserve, Kent, England

I see the male first. He saunters towards the edge of the enclosure, nonchalantly. Reaching the wire, he sniffs at the grass and then, standing still, stares straight ahead at the small crowd of admirers looking right back at him. He has the confidence of a champion boxer, as if he were waiting patiently by the side of a ring for the sound of the bell, so he can show us all what he's made of.

From his golden face, pale amber eyes, unblinking, keep a constant gaze on us. Around his eyes, his fur is a lighter shade of gold than the rest of his body. It looks as if someone has puffed his face with iridescent powder, dusting generously around his eyes to accentuate their bright light. Standing there, bathed in the sunshine of a fabulous summer's day, he has the face of a sun king.

But it's his enormous mane, voluminous and shiny, that steals the show. It starts above his head, all tawny and golden, hot ginger flecks shimmering here and there. As it passes over the back of his head, it turns darker, taking on mahogany tones. Continuing down over his shoulders it falls like a cape, eventually covering his stomach, creating a luxuriant gold-threaded belly beard.

His wide and powerful shoulders are hidden under his mane, but as he turns away from the wire the rest of his body, toned and taut, comes into view. He's broad at the shoulders, but tapers in thinly, tight as a snake, around his back legs. As he moves away swiftly, the black-tufted tip of his serpentine tail cracks the air like a whip.

He returns quickly, eager to face his crowd, but this time he is not alone. Leading the way is the female Barbary lion with which he is paired. Strong as Boudicca, she walks with queenly dignity. She doesn't have the glamorous and extravagant mane of the male, but she has presence, quietly powerful, understatedly noble.

She holds her head high, perfectly balanced and still, as if waiting for a crown to be placed upon it. A soft diamond-shaped pink patch sits in the middle of her nose and around her almond-shaped eyes, honey-hued fur covers her face.

Together they make the perfect feline fairy-tale couple. Even when the female snarls fierce and deep from her velveteen muzzle, neither has a whisker out of place. Enjoying the heat of the sun, they nuzzle, their little under-beards showing. In the sunshine the lions gleam brightly, as does the hope that they will bring cubs, laced with Barbary genes, into the world.

Milo and Ruti seem happy enough. He has his Spanish queen and she the most handsome of lion beaus. As I turn to leave, Ruti flicks her tail upwards. Its black-tufted tip disappears, lost in the darkness of Milo's mane as he lets out a mighty roar, just like his wild ancestors once did in the mountains of Morocco.

A while after my visit to the reserve, I learn that Ruti died before the pair produced any cubs – an indicator of the precarious nature of captive breeding large animals that aren't easily able to choose their own mates, and when bloodlines must be managed through the transfer of animals between geographically distant zoos. Finding a new mate for Milo is challenging. A new female must be carefully selected. Within her projected lifetime she needs to be able to breed successfully and preserve genetic diversity. Finding a mate that fits the bill is difficult, especially when some zoos are reluctant to share 'their' lions.

Since their extinction from the north of the continent, lions are now only found south of the Sahara. In West Africa, they hang on by the most fragile of whiskers. They are the least protected and most endangered lions on the planet.

In January 2014, the results of a six-year study assessing the number of lions in West Africa called 'The Lion in West Africa Is Critically Endangered' were published in *PLOS ONE*. It showed that just 406 lions remained in the entire region, all of them facing extinction. It also revealed that these few remaining lions could be found in only five countries, compared with fourteen some fifteen to twenty years ago.

Led by Philipp Henschel, a Panthera lion programme survey co-ordinator based in West Africa, one of the study's prime objectives was to assess how many lions survived in the region. Historically, the region, which has extremely low rates of investment in conservation, has been under-surveyed, with only a limited amount of data about lion numbers and their distribution available. Talking to *National Geographic* about the lack of research prior to his survey, Henschel said: 'It was really not known that the status of the lion was so dire in West Africa … In many countries it was not known that there were no more lions in those areas because there had been no funding to conduct surveys.'

To collect up-to-date and region-wide data, the survey, carried out between October 2006 and May 2012, looked for lions in eleven West African countries (Benin, Burkina Faso, Côte d'Ivoire, Ghana, Guinea-Bissau, Guinea, Mali, Niger, Nigeria, Senegal and Togo), believed to be lion range states. A number of counting methods were used, including call-up, track and aerial surveys, lion track and spoor counting, remote camera trapping and interviewing conservationists who worked locally with lions.

Within these eleven countries, twenty-one areas where lions were thought to still survive were surveyed. Of these, thirteen were relatively large protected areas, more than 498 square kilometres each in size. The remaining eight areas were small and unprotected.

Only four small and isolated populations were found in five of the eleven countries surveyed. The five countries were Benin, Burkina Faso, Niger, Nigeria and Senegal.

Henschel estimated that the largest population of 356 lions was at the W-Arly-Pendjari (WAP) complex, a vast protected area covering more than 29,000 square kilometres, crossing the borders of Benin, Burkina Faso and Niger.

Outside of the WAP complex, in the rest of the three border countries, no lions were found. This wasn't always the case. In 2012, Riggio et al.'s 'The Size of Savannah Africa: A Lion's View' survey estimated that forty lions lived outside the complex in Benin. And according to a quote cited by Philippe Chardonnet in his 2002 survey: '[In 1956] the lion used to be present everywhere in Burkina Faso'. Chardonnet also noted that at the time of compiling his 2002 results, despite prey being available lions were gone from western and central Burkina Faso.

Likewise, Niger's lions also used to survive in the country outside of the WAP area. In his 2002 study, Chardonnet quotes survey findings from 1939 that stated Nigerien lions were 'especially widespread ... near the Mali border' and 'common in the Air Massif' in the north of the country.

West Africa's other three surviving populations in Nigeria and Senegal are catastrophically tiny. In Nigeria, thirty-two lions were counted at Kainji Lake National Park by Riggio et al. Just two were found at the Yankari Game Reserve. In 2009, forty-four lions had been counted at Kainji Lake. Two years later, in 2011, the number was down by ten to thirty-four.

Up in Senegal, just sixteen lions were recorded by Henschel at the Niokolo-Koba National Park, in the south-east of the country. In 2002, Chardonnet estimated that 125 lived in the park, suggesting a loss of 84 lions in just 12 years.

As well as reporting on the numbers of surviving lions in West Africa, Henschel's study also estimated that of the region's 406 last lions, only 250 were sexually mature – a number too small for the population to breed itself out of extinction. Scientists have a phrase for animals with such small numbers. They call them the 'living dead', a sentiment echoed by Pieter Kat, biologist and lion expert, when commenting on the plight of West African lions to *The Guardian* in 2013: 'The populations are declining so quickly, as a biologist I would say that in a country like Nigeria where there are only 34 lions, they are already extinct. It's almost impossible to build up a population from such a small number.'

West Africa's last lions now live in just 1.1 per cent of their historic range; a range that once stretched from Senegal down to Nigeria. A huge backyard, some 2.4 million square kilometres in size, now reduced to just 4,800 square kilometres. This extreme decline in range reflects their decline in numbers. Their predicament is dire; their fall calamitous for the region.

Although Henschel's call for the region's lions to be listed as critically endangered was answered by the IUCN in 2015, lions in West Africa are still walking the same path to extinction followed by their cousins in the north less than seventy years ago.

Further down the continent, in the heart of Africa, lions also struggle for survival. Just a few thousand remain in this region. Most are found in Angola and the Central African Republic, while others survive in the Democratic Republic of Congo, Cameroon and Chad. In the Republic of Congo, the lions are all gone. And, until the summer of 2015 when a single lion was captured on a camera trap, it was believed they had vanished from Gabon too.

The most recent estimate for the total number of lions in the region is cited as 2,419 in Riggio et al.'s 2012 study 'The Size of Savannah Africa: A Lion's View'. It pinpoints just two large populations in the region.

The largest, with 1,905 lions, is in the south-east of Angola. The second-largest population, comprising 1,244 individuals, is in the east of the Central African Republic. Despite their size, neither of these populations are considered a lion stronghold because their numbers, although relatively large, are still in decline.

In Angola's north-west at Kissama-Mumbondo, Riggio et al. estimated that ten lions remain, with another fifty-five hanging on at Bocoio-Camucuio in the west. A larger population of 455 lions, shared with bordering Namibia, survives at Etosha-Kunene in the south.

Gathering accurate data for lions in Angola is regarded as particularly challenging. Chardonnet, who estimated that 649 lions survived in the country in his 2002 survey, noted: 'The status of lions in Angola is poorly understood ... Crude estimates for some regions of the country do exist, but these, however, were not based on scientific surveys.'

In fact, throughout Central Africa, as in the West, historic lion data is considered problematic. In 2012, for example, Riggio et al. described it as 'unreliable', and Chardonnet in

2002 described it as a region where numbers are 'poorly known'.

In the Central African Republic, Riggio et al. suggested that lions remained in two other areas at Bozoum and Nana-Barya in the east, where tiny populations of just four lions each cling on. In 2002, Chardonnet estimated that 986 lions in total remained in the Central African Republic, compared with a 2010 survey led by Pascal Mésochina that estimated a possible 1,289 lions living in the country. As in Angola, Chardonnet in his 2002 survey noted that in the Central African Republic, 'very few lion studies have been carried out', and cited the observations made in 1958 that lions that were once abundant were now a rare sight.

Up in the north-west of the region in Cameroon, Riggio et al. estimated that a total of 217 lions, living in 2 populations, remained in the country. In 2002, Chardonnet put the total number of lions at 415, noting there had been evidence of larger lion numbers in the past living in the country.

In 2012 Riggio et al. reported that Cameroon's largest remaining group of 200 lions survived in the north-east of the country at the Benoue Complex. The country's seventeen other lions are at Waza National Park, where they share a border with Nigeria in the far north. In the 1960s, a survey conducted at Waza estimated the park to be home to 200 lions.

In Chad, Riggio et al. in 2012 suggested that 400 lions survived in the south-east of the country, but also commented: 'The validity of this estimate is questionable given the lack of scientific research conducted in this region.' Ten years earlier, Chardonnet estimated that a possible 520 lions could be found in Chad.

Just 175 lions were estimated as surviving in the Democratic Republic of Congo in 1 population in the north by Riggio et al. In 2002, Chardonnet concluded that 606 lions roamed the country. Eight years later in 2010, lions in the national parks at Kundelungu and Upemba, both in the southern Democratic Republic of Congo, were reported as gone.

In the Republic of Congo next door, lions have vanished altogether. In 2002, Chardonnet estimated that a total of sixty lions survived in the country – forty at Odzala National Park in the west and twenty at Batéké Plateau National Park in the south-west. In 2006, the IUCN suggested that perhaps fifty lions remained in the republic. Four years later in 2010, a survey conducted by Henschel et al. commented that because 'no material evidence for the species had been produced in the last 15 years', it was 'reasonable to assume that resident populations are extirpated'.

Likewise, lions are assumed to be extinct in neighbouring Gabon. There, the last suggestion that any lions remained in the country was by Chardonnet in 2002 when he estimated that twenty lions, shared with the Republic of Congo, survived at the Batéké Plateau National Park in the south-east.

However, in 2015, fresh hope that lions weren't gone for ever came glimmering through a Gabonese forest. Early that year, CNN reported that a lone male lion had been caught on camera in the Batéké Plateau National Park. This was the first time a lion had been seen in Gabon since 1996.

Footage of the lion had been unintentionally captured on film by wildlife researchers working in the park studying primates. After watching the film, Henschel, who is based in Gabon, expressed his shock, since he believed lions to be extinct in the region since at least 2001, telling CNN: 'This footage is truly unexpected, and yet wonderful proof that life for the lions of Gabon and the region still remains a possibility.'

Henschel believed that the lion probably travelled almost 250 kilometres from the Democratic Republic of Congo into Gabon, quite possibly looking for a female. Other evidence also suggests that the lion had been living in the park since November 2014, indicating that the lion may have made a base for itself in the country.

There's footage of Gabon's lone lion online. The night-time clip is short, around thirty seconds long, and shows a young lion with a dark mane wandering along a path used more often by the park's forest elephants. The lion appears strong and looks well-nourished. He looks confident too, roaming the forest he shares with western lowland gorillas and chimpanzees, completely unaware of the hope his presence there has inspired.

In September 2016, almost two years on, the same lion was unexpectedly caught on film again. Just as before, he was filmed at night in a similar location, still looking healthy, but this time he was much closer to the camera, providing a close-up of his short dark mane, typical of male lions in Central and West Africa.

Despite the optimism that Gabon's solitary big cat generates, the outlook for lions in Central Africa and those in West Africa is bleak. Most countries in the two regions have lost their lions altogether and those that remain risk their lives every day in some of the continent's most poorly protected areas.

Many of these populations are small and isolated, living too far away from other lions that might demographically or genetically save them from the brink of extinction. Put simply, lions in the west of the continent are flatlining and those in Central Africa are just moments away from doing the same.

East Africa, most particularly Kenya, is intrinsically associated with lions. It's their evolutionary birthplace some 3 million years ago, it's where Elsa the *Born Free* lioness returned to the wild and where *Out of Africa* was filmed. Jonathan Scott's *Big Cat Diary* played out here too, and it's the fictional home of Disney's Simba, the Lion King. In the public imagination, it's a region where lions are still kings.

In 2015, the publication of Bauer et al.'s 'Lion Populations Are Declining Rapidly Across Africa, Except in Intensively Managed Areas' challenged this misconception. It revealed that in all of East Africa just one large population of lions, located in the Serengeti in northern Tanzania, had not experienced a fall in numbers, and is the only population in the region that can be considered to be stable.

Home to 40 per cent of Africa's lions, Tanzania has by far the largest number of lions on the continent. In 2012, Riggio et al. estimated that 16,984 lions, in 9 populations, survived in the country. Of Africa's ten lion strongholds, four of them are found in Tanzania.

The largest population of 7,644 lions is at the Selous Game Reserve in the south. They're followed by 3,779 lions at Ruaha-Rungwa, also in the south. The Serengeti-Mara population described by Bauer and van der Merwe as 'the only large population surveyed that is not declining' comprises 3,673 lions, which it shares with Kenya.

The 880 lions at Tsavo-Mkomazi, in the west, are also shared with Kenya. Tanzania's remaining five populations are very small – with just fourteen lions counted by Riggio et al. at the Mpanga Kipengere Game Reserve in the southern highlands of the country.

A decade earlier, Chardonnet estimated that 3,896 lions might remain at Serengeti, with a possible 3,360 living at Ruaha-Rungwa. He estimated the smallest population of eighteen lions to be at the Mahale Mountains National Park in the north-west of the country.

In neighbouring Kenya, Riggio et al. (2012) suggested a total of 5,644 lions survived in 6 populations, the largest being the 3,673 Serengeti-Mara population shared with Tanzania. Kenya also shares another population of 880 lions with Tanzania at Tsavo-Mkomazi in the south. The 750 lions at the Arawale-Bush Bush area in the north-east are shared with Somalia.

Kenya's smallest population of fewer than thirty lions was counted in June 2012 at the Nairobi National Park, just outside the country's capital. Conducted by the Friends of Nairobi National Park, a conservation society affiliated with the park, it noted: 'As of 25 June 2012, Nairobi Park is home to 29 adult, sub-adult and juvenile lions. In addition, there are 11 known cubs less than 6 months of age.'

However, in 2002, Chardonnet estimated that 2,925 lions remained in Kenya, comprising 18 populations of which only 2 were home to more than 500 lions each. Chardonnet suggested that the biggest population of 547 lions was at the Masai Mara National Reserve and surrounding areas and the smallest, of just 9, at the Hells Gate National Park and the former Kedong Ranch area, north-west of Nairobi.

In 2012, Riggio et al. suggested that just two populations remain in Somalia. The largest group of 750 are lions shared with Kenya at the Arawale-Bush region, in the south-west. The only other population, of 175 lions, is located at Arboweerow-Alafuuto in the south of the country. In 2002, Chardonnet estimated that a population of 128 lions remained at the Swamp National Park in the northern Great Rift Valley. By 2012 they were all gone.

In the north of East Africa up in Ethiopia, Riggio et al. estimated that the country had seven populations remaining. The largest group of 500 lions was located at Boma-Gambella in the west of the country, shared with South Sudan. At South Omo in the south-west, some 200 lions were estimated to survive.

The smallest population of just ten was counted at Nechisar in the Great Rift Valley, in the south-west. In total, Riggio et al. suggested that Ethiopia was home to around 1,000 lions in total. In 2002, Chardonnet had estimated that a total of 2,147 lions lived there.

However, in 2016 a new population of between 100 and 200 lions was discovered at the Alatash National Park in the north of the country, bordering Sudan's Dinder National Park. About the discovery, Bauer, who was working in the region at the time, told *New Scientist* magazine that he had spent much of his career removing lion populations from his distribution map, but now – since this sighting – it was 'the first and probably the last time that [he was] putting a new one up there'. Before this discovery, lions had been considered regionally extinct in Sudan since 2011.

In South Sudan, Riggio et al. noted three small populations, all of them shared. The largest group of 500

lions was shared with Ethiopia in the east at Boma-Gambella
and another 375 lions shared with the Central African
Republic in the south-west. In the south-east at Kidepo
Valley, the smallest population of 182 lions was shared with
Uganda. In 2002, Chardonnet counted six populations in the
country, half of them having just sixteen members each, all
of which are now gone.

As well as the population of 182 lions shared with South
Sudan, Riggio et al. noted that Uganda is home to 4 other
populations. The largest of 210 is shared with the Democratic
Republic of Congo at Greater Virunga in the Albertine Rift.
The smallest is a tiny population of just three lions at Lake
Mburo National Park in the west.

A 2013 survey counting both lions and hyenas at the
Queen Elizabeth National Park in the west of Uganda
counted 144 lions and commented: 'It is estimated that lion
numbers have declined by 30 per cent in this protected area
since the late 1990s and there are increasing concerns for the
long-term viability of both species in Uganda'. In 2002,
Chardonnet had estimated a population of 224 lions living at
the park.

Since 2011, lions have been considered regionally extinct
in three East African countries – Burundi, Djibouti and
Eritrea. The only recent sighting in Burundi have been of
lions occasionally entering the country from bordering
Tanzania. Likewise, Djibouti receives the odd lion visitor
from neighbouring Ethiopia. About Eritrea's lions,
Chardonnet in 2002 commented: 'The lion was probably
present in Eritrea during recent times, from where we have
no record now.'

If it wasn't for the reintroduction of seven lions from
southern Africa to the Akagera National Park in the north-
east in 2015, Rwanda would be the fourth East African
country with no lions. In 2002, Chardonnet estimated that
forty-five lions lived at Akagera. A decade earlier, that
number had been as high as 250, but by 2010 'no signs of lion
activity have been recorded, suggesting that lion populations

have been recently extirpated'. The population is growing, with nineteen lions now found at the park.

In all of East Africa, the lions at Serengeti are the only lions currently likely to maintain their population size. Bauer et al. in 2015 revealed that all other populations from Sudan to Kenya and the other lions in Tanzania have – unless the current trend is bucked – a 37 per cent chance of falling again by half over the next twenty years. Smaller populations like the fifty at Tsavo East and Tsavo West national parks in Kenya will likely vanish altogether.

Down in the south of the continent, the outlook for the region's 12,000 or so lions is more promising. Of the ten lion strongholds in the continent, six of them are found in southern Africa. In Botswana, Namibia, South Africa and Zimbabwe, lion populations are stable or increasing overall. Lesotho is the only country, out of the ten[*] in the south, to no longer have lions.

South Africa is the best place to be a lion. It's the only country in the continent where all lion populations have remained stable or gained in number. This wasn't always the case, though. Before the nineteenth century, lions were considered abundant and roamed most of the country. During the 1800s, an estimated 1,492 lions were lost from the Northern Cape and north of the Orange River and were gone from the Eastern Cape by 1879. By the late 1800s and early 1900s, South Africa's lion population had been decimated. And by the 1950s, the lions that remained were found only in the country's newly established national parks.

However, where lions once lost ground they have now returned, largely because of the proliferation of private game reserves since the 1990s, where lions have been

[*] Angola, Botswana, Lesotho, Malawi, Mozambique, Namibia, South Africa, Swaziland, Zambia and Zimbabwe.

successfully reintroduced. Lions are now found, usually as fenced-in populations, in all of South Africa's provinces bar the Free State.

When Riggio et al. estimated the number of lions in South Africa, they only included lions that roam the country's three unfenced transfrontier parks, where wildlife roams freely between protected areas that span international borders. A total of 2,311 lions were estimated to be in residence at the Great Limpopo Transfrontier Park, which includes the Kruger National Park and straddles Mozambique in the east and Zimbabwe in the north.

A smaller population of 800 is shared with Botswana in the north at the Kgalagadi Transfrontier Park, which takes in South Africa's Kalahari Gemsbok National Park and Botswana's Gemsbok National Park. At the Greater Mapungubwe Transfrontier Conservation Area in the north-east, which includes South Africa's Mapungubwe National Park, Botswana's Northern Tuli Game Reserve and Zimbabwe's Sentinel Ranch, Nottingham Estate and the Tuli Circle Safari Area, twenty-five lions are estimated to roam.

Regarding the number of lions at small and fenced-in reserves, Susan Miller et al. in 2016 noted that the number of small reserves with lions increased from just 1 in 1990 to at least 45 in 2013, corresponding to a growth in numbers from around 10 lions to 500. Miller also comments that if the count for South Africa's lions only included those in transfrontier parks and excluded subpopulations in small and fenced reserves, then the country's lions would 'technically qualify for Near Threatened' status, but because overall numbers are stable or rising, and since the 'reintroduced subpopulations on small reserves qualify as wild and free roaming, a Least Concern listing is most appropriate'.

Considering how well lions are doing in South Africa compared with the rest of the continent, it's not surprising that most of the lions I've seen have been here. The first lions I saw outside of a zoo were at the Madikwe Game Reserve up in the north. In the Eastern Cape at the Shamwari Game

Reserve I watched a giraffe crumple to the ground, brought down by a hunting pride. But it's a family of rare white lions that I remember most vividly.

Sanbona Wildlife Reserve, Western Cape, South Africa
The mountain range around me glitters brightly as if dusted with diamonds. These mountains, sprinkled with deposits of quartz, are the sparkling boundaries of the Sanbona Wildlife Reserve in the heart of South Africa's Western Cape.

I'm writing a piece about white lions that have been released here after spending their lives in captivity. Paul, a Sanbona wildlife professional, is driving me through the reserve in hope of a sighting. As we move into the plains, dusty rocky outcrops and unyielding boulders make way for greener vegetation.

Zebras and antelopes stare as we pass by. An ostrich sits down on her sprawly nest, protecting her egg like a giant pearl. Paul stops the jeep, scans the horizon and starts to smile. Following his gaze, I soon realise why. He has seen the white lions.

Three cubs, about ten months old, tumble around their father. Against the green of the plains, nourished by recent rains, the colour of the lions is startling. With pale blue eyes and shining coats the colour of vanilla ice cream, they are unlikely looking beasts of the bush.

The cubs start playing pretend-predators. Limbs taut with muscle, they pounce and leap, instinctively preparing for life outside the pride. As they play, their father watches over them. Sitting haughtily on his haunches, still as an alabaster statue, he appears unearthly, almost spectral. His ethereal gaze makes it easy to understand how white lions have long been the stuff of African legend.

For centuries, from Egypt in the north to Namibia in the south, stories about white lions have been told all over the continent. In some, the lions have been cast as animal Houdinis, able to appear and disappear magically at will. In others, they are blessed with supernatural powers or revered as gods. In Timbavati, almost 2,000 kilometres from here, the Shangaan people tell tales of white lions cast as the holy messengers from another universe.

Anatomically and behaviourally, white lions are just like tawny lions. They're not albinos, though. Rather, they are the result of a gene that prevents them from being golden in colour. Because this gene is recessive, white lions appear unexpectedly and often generations apart. If one lion carrying the gene mates successfully, there is just a 1 in 5,000 chance that a cub will be white.

This rare possibility that they'll appear helps to explain why the white lions of African folklore are depicted as animals that can disappear at will. White lions are random, beautiful quirks of nature, carrying genes that predispose them to the mythological.

The cubs continue to enchant with their whiskery, white-cat magic, until the female decides it is time for them to leave. In true follow-my-leader style, the cubs follow their mother and are gone, once again vanished into thin air.

While South Africa today might be a good place to be a lion, this hasn't always been the case. For the Cape lion that once roamed the country's most southerly points, South Africa from the mid-1600s became a killing field. Once seen frequently around Cape Town, Cape lions became a less regular sight. A couple of centuries later, they were only seen further north in the semi-desert region of the Karoo and to the edges of Uitenhage in the east. By the 1860s, the Cape lion was extinct from the region. Not a tail, not a whisker or claw remained of Africa's largest and heaviest lion. Leaving the planet quietly, it was the first of Africa's lions to be lost from the continent for ever.

In 1876, Emil Holub, an explorer from Czechoslovakia (now the Czech Republic) was on an expedition in South Africa. During his travels in the Cape, the explorer came across two, by this time now rare, Cape lion cubs – one male, one female – for sale. Holub bought the male as a travelling companion and, impressed by the cub's emerging majesty, called the young lion Prince.

So Prince could travel comfortably, Holub adapted his wagon, creating a cage-like area at the back. Most nights, Holub and the lion rested together, Prince sleeping like a devoted dog at his master's feet. The pair became inseparable.

When the time came for Holub to return to Czechoslovakia, Prince travelled with him. Safely back in Europe, Holub made Prince a new cage, spending time in it, sitting down and talking to his now huge pet whenever he could.

After a while, Holub set sail for Africa to lead another expedition. While he was away, Holub's servants were tasked with caring for Prince. When Holub returned, however, he found his beloved Prince dead. His servants, too frightened to go inside the lion's cage to clean it out, had left Prince's cage so dirty that the gases from the lion's own excrement fatally and un-majestically poisoned him at just two years of age.

Distraught, Holub wrote in his diaries: 'I had jackals, who repeatedly ran away and came back again, gerbils, who left as infants, but none was able to make me replace my Prince.' Unable to say a complete goodbye to his lion, Holub had Prince stuffed. Prince is still with us, now as an exhibit in the Emil Holub Museum in Holice in the Czech Republic.

In 1859, seventeen years before Prince set sail for Europe, another male Cape lion was shipped from South Africa, bound for English shores. Coming into London up the Thames, the lion, later known as Nelson, was offloaded at Tobacco Docks in Wapping, part of a cargo that included other live African animals, as well as the furs, skins and ivory of those that were not so lucky.

There to greet Nelson was Charles Jamrach, a merchant who specialised in importing wild animals for his commercial menagerie, making a living selling his exotic wares to an often equally exotic clientele, including the poet Dante Gabriel Rossetti, the circus owner P. T. Barnum and numerous aristocrats and celebrities of the day.

From the menagerie, possibly via a stint in a travelling circus touring Europe, Nelson spent the last thirteen years of his life at London Zoo, where he was admired by Londoners, agog at having such a fabulous creature in their midst.

When Nelson died in 1872, he was acquired, stuffed and mounted by Rowland Ward, a famous taxidermy firm. He was then bought by B. J. Whitaker, a collector of natural history materials from all over the globe, and transported to Whitaker's private museum at Hesley Hall in Yorkshire in the north of England. In 1973, Nelson was given to the Clifton Park Museum in Rotherham.

Apart from paintings and old photographs, the only way to get a really good look at a Cape lion is to find a museum specimen, one like Nelson or Prince. This isn't as simple as you might think. When the Dutch author L. C. Rookmaaker compiled the writings and discoveries of early zoological explorers in southern Africa in 1989, he wrote that few Cape lion specimens could be found in museums and that for 'a long time, it was assumed that there was only one, a mounted male in the Junior United Services Club, now in the British Museum'.

The specimen Rookmaaker refers to is now at the Natural History Museum in London, England. This lion, male and black-maned, was shot in 1836 by Captain Copland-Crawford of the British Royal Artillery by the side of the Orange River in Colesberg, South Africa. When first killed, he had been sent all the way from the Cape to London, where he was stuffed and mounted by Gerrard & Co., a well-known firm of taxidermists.

There are a pair, male and female, at the Museum Wiesbaden in central-western Germany and another pair at the State Museum of Natural History in Stuttgart in the south-west of the country. France has a male Cape lion (and a Barbary lion) on display in the Room of Endangered and Extinct Species at the Natural History Museum in Paris. Another is at the National Museum of Natural History in Leiden in the mid-west of the Netherlands. There are also a

handful of skulls in the natural history museums of Port
Elizabeth in South Africa and Copenhagen in Denmark.

Although historically Cape lions were kept in menageries
and zoos around the world, they were bred with captive lions
from all over Africa and became hybridised. The last Cape
lion kept in captivity is believed to be a male lion held at the
Zoological Gardens in Paris in the 1860s.

Tracing lions with Cape lion ancestors is problematic.
Few breeding records exist and there aren't any stud books
similar to the one that identifies possible descendants of
Barbary lions. Furthermore, recent DNA tests on Cape lion
remains have revealed that even though Cape lions were
noticeably bigger and darker than other African lions, they're
not genetically distinct from other lions in southern Africa.
This means that there's no way to test a lion for Cape lion
heritage.

Although it's now considered pretty much impossible to
revive the Cape lion, in 2000 a zoo director from South
Africa called John Spence travelled to Siberia from South
Africa to do just this.

Spence had a lifelong fascination with Cape lions. Stories
of the lions had intrigued him since childhood, especially the
exciting accounts of enormous Cape lions scaling the walls of
a Cape Town military fort built in the 1600s. He wanted
passionately to reintroduce Cape lions to the Cape and
believed that if he could find descendants living in captivity
he might be able to do this.

Spence started looking for such lions in the 1970s. He
searched zoos and circuses around the world, as far afield as
Singapore and the United States, hoping to find animals that
looked so much like Cape lions that they might have Cape
lion ancestors. Thirty years on, a friend sent Spence images
of a lion living in Novosibirsk Zoo in Siberia, that was the
spit and image of a Cape lion – incredibly large, long-backed
and with a super-sized dark mane. The lion was called Simon
and had been bought from a travelling circus by the zoo
in 1961.

When Spence saw the photograph, he told *National Geographic*: 'Every hair on my body stood upright, including my neck and everywhere else!' This lion looked so much like a Cape lion that he must surely have Cape lion heritage, thought Spence. With assistance from Vienna Zoo, Spence and his wife Lorraine travelled to meet Simon at the Russian zoo. By the end of the trip, Spence had become the proud owner of two of Simon's cubs, named Olga and Rustislav.

The cubs flew back with the Spences and were homed at Tygerberg Zoo in the northern suburbs of Cape Town. The idea was that once the cubs reached sexual maturity, they would be used to start a Cape lion breeding programme.

However, this never happened. Spence died in 2010 and Tygerberg Zoo closed down two years later in 2012. It's believed that Olga and Rustislav were transferred to a lion sanctuary in Drakenstein to live out their days, taking any possible Cape lion ancestry with them.

In Botswana, Riggio et al. estimated in 2012 that four populations of lions survived. The largest group of 2,300 lions was counted at the Okavango-Hwange lion area, covering much of northern Botswana and Zimbabwe's Hwange National Park. Although this population is relatively large, it's the only population in southern Africa thought to be declining.

Another 100 lions are shared with South Africa: 800 at the Kgalagadi Transfrontier Park and a further 25 lions at the Greater Mapungubwe Transfrontier Conservation Area. At Xai-Xai in the east, a population of seventy-five exists. In 2002, Chardonnet estimated that 3,217 lions were present in Botswana, compared to Riggio et al.'s estimation of 3,200 a decade later.

In Zambia, Riggio et al. in 2012 counted six lion populations, the two largest of which are shared with neighbouring Zimbabwe and Malawi. An estimated 755 are at the Mid-Zambezi lion area covering a large part of southern Africa and northern Zimbabwe, and a further 574 border-crossing lions are shared with Malawi at the Luangwa lion area in the south-east. The remaining 4 populations range in number from 386 at the Kafue National Park in the west down to just 4 at Liuwa Plains, near the Angolan border.

Up in Angola, Riggio et al. estimated that four lion populations remained. The largest group of 1,905 lions is in the south-east of the country. At the Etosha-Kunene lion area in the south, 455 lions are shared with Namibia. In the west at Bocoio-Camucuio, there are fifty-five lions. The smallest population of fewer than ten lions is at Kissama-Mumbondo in the north.

In 2002, Chardonnet estimated there were 649 lions in all of Angola, but quoted a personal communication from wildlife professional Wouter van Hoven outlining his concern with the accuracy of lion counts in Angola: 'Crude estimates for some regions of the country do exist, but these, however, are not based on scientific surveys.'

Towards the west of the region, in Malawi, Riggio et al. estimated the numbers of the country's three lion populations. The largest, of 574, a shared population with Zambia, is in the north-east at Vwaza Marsh Game Reserve. Just eighteen remain at Nkhotakota Wildlife Reserve in central Malawi, and six, also in central Malawi, survive at Kasungu National Park.

In 2013, three lions were translocated from South Africa to the Majete Wildlife Reserve in the south, where two cubs were born in 2016. In 2002, Chardonnet estimated just twenty-five lions to be in Malawi, ten of them at Nkhotakota in central Malawi and five at Liwonde in the south.

Next door in Zimbabwe, Riggio et al. counted five populations. The largest population of 2,311 lions is shared with South Africa and Mozambique at the Great Limpopo Transfrontier Park in the north-west. Another good-sized population of 2,300 lions is found at the Okavango-Hwange lion area, and is shared with Botswana.

A smaller but still significant shared population of 755 roams at the Mid-Zambezi lion area, where they cross into

Mozambique and Zambia. At the Greater Mapungubwe Transfrontier Conservation Area, twenty-five lions are shared with Botswana and South Africa.

In the south-east at Gonarezhou National Park, a 2010 survey counted twenty-three lions. Four years later another survey estimated that thirty-three lions roamed the park. Numbers of lions have also increased at the Bubye lion area in the south, where 200 lions are estimated to live after 17 lions were initially reintroduced into the area in 1999.

In the north-east of the region in Mozambique, Riggio et al. recognised six populations. The largest group comprises 2,311 lions at the Great Limpopo Transfrontier Park, crossing borders with Zimbabwe and South Africa.

The second-largest population of 1,573 lions is found at a transfrontier conservation area linking the country's Niassa Reserve in the north with Tanzania's Lukwika-Lumesule Game Reserve. In 2002, Chardonnet estimated that 500 lions were at the reserve.

It's estimated that the number of lions has increased at Niassa by around 250 per cent since 1993. Conservationists have suggested that during this time, lions at the reserve benefitted partly from the mass slaughter of elephants by poachers, enabling them to thrive on the large amounts of elephant meat left behind by poachers only interested in the animals' tusks.

Mozambique's third population of 755 lions is shared with Zambia and Zimbabwe at the Mid-Zambezi lion area. The remaining three other lion groups are smaller and centrally located. At Gorongosa-Marromeu, there are an estimated 229 lions, an additional 59 at Tete province, with the smallest population of 45 found at Gile.

Namibia, according to Riggio et al.'s 2012 survey, is home to two populations. The largest group of 455 lions is shared with Angola in the north-west at Etosha-Kunene. The country's other population comprises 150 lions at the Khaudom-Caprivi lion area in the north-east. A decade earlier, Chardonnet estimated a total of 417 lions at Etosha-Kunene and 274 at

Khaudom–Caprivi, a total of 691 lions, compared with Riggio et al.'s 605.

Just two countries in southern Africa have lost all their lions. Lesotho lost its lions first, most likely in the late 1800s. An entry in *Encyclopaedia Britannica* remarks: 'Hunting and deforestation have mostly eliminated the populations of large mammals; the last lion was killed in the 1870s.' And a 1996 report for the IUCN notes: 'Lions are believed to be extinct or practically so in … Lesotho.'

Lions became extinct in Swaziland in the 1950s, when the last lion in the land was reportedly seen by the king, Sobhuza II. However, some thirty years later, three lions from South Africa's Kruger National Park were introduced to the country's Hlane Royal National Park, in the north-east. In 2002, Chardonnet estimated that nineteen lions were at large in the park and that Swaziland was once again a lion range state.

Just as lion kingdoms in Africa have collapsed, so too have their outposts beyond the continent. The lions that left the motherland and once had a huge historic range, including northern Greece, the southern Balkans, much of the Middle East and South West Asia, have lost incredible ground. It's a range they now barely inhabit.

By the close of the first century, lions were gone from Greece. Around 900 years later, Georgia, Armenia and Azerbaijan said goodbye to their lions, too. During the Third Crusade, in the late 1110s, lions disappeared from Jordan. Three hundred years later, Palestine's lions did the same.

In Pakistan, lions hung on until 1842, when the last one was killed near Kot Diji at Sindh, in the south-east. However, in February 1935, Philip Dumas, a British admiral with the Royal Navy, maintained that he and two travelling companions had seen a lion from their railway carriage at the Bolān Pass in West Pakistan. Dumas's sighting was disputed

Map 3: Historic and current range of Asiatic lions

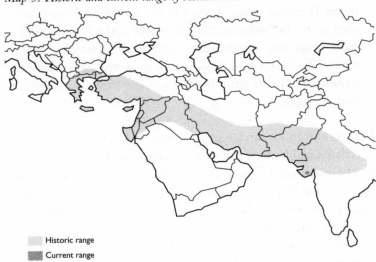

Historic range
Current range

by many, prompting him to write a letter to *The Field*, a British country sports magazine, confirming that he and his companions had clearly seen a lion 'eating a goat' that was very large, 'very stocky, [and] light tawny in colour', and that not 'one of the three had the slightest doubt of what [they] had seen'.

During the late 1880s, it was Turkey's turn to lose its lions, prompting Alfred Pease, a British politician and early settler in Kenya who wrote a number of books about his travels, including *The Book of the Lion*, to visit Turkey in search of lions. Unable to find any, he despondently wrote in 1891: 'The lion does not occur anymore in Asia Minor.'

In Iraq, the last lion is said to have roared in 1918, when it was killed by a Turkish statesman at the lower reaches of the Tigris River. Fifty years or so earlier, lions had been regularly spotted by British archaeologists excavating ancient Assyrian ruins. In 1849, for example, Sir Henry Layard, who was digging at the Assyrian city of Nineveh on the outskirts of Mosul in northern Iraq, wrote in his book called *Early Adventures in Persia, Syria, and Babylonia*:

The lion is frequently met with on the banks of the Tigris, below Baghdad, rarely above. On the Euphrates it has been seen … In the Sinjar, and on the banks of the Kabour, they were frequently caught. They abound in Khujzistan, the ancient Susiana: I have frequently seen 3 or 4 together and have hunted them with the chiefs of the tribes that inhabit that province.

During the 1850s, archaeologist and explorer William Loftus excavated at Uruk in the south and encountered many lions. In the accounts of his travels, Loftus wrote about a lioness that frequently visited his camp, taking and killing the camp's dogs until just the favourite called Toga remained. Loftus recounts how Toga was then also carried off by the 'dog devourer', causing much misery the following day.

In neighbouring Iran, lions persisted in small and isolated populations until the 1960s. Their decline started in the early 1900s. In 1908, Sir Arnold Wilson, a British officer, recalled walking near Ahvaz, in the south-west, with a Persian companion when a lion came into view. His walking chum wished to take aim at the lion, but Wilson persuaded him otherwise. Later on, he noted his concern for the country's lions in his diary: 'There are so few left … It is a pity the Persian Government do not create a sanctuary for them, seeing that the lion is the national emblem of Persia.'

By the 1940s, Iranian lions had largely disappeared, with just a few sporadic sightings by railway engineers recorded during the Second World War. Their final exit was in 1963 when the country's last lion, a female, was shot and killed in a cave at Fars Province in the south of the country. Her four cubs were taken as trophies and then placed in captivity.

Country by country, the lions that migrated from Africa on an odyssey spanning lands from the tip of southern Europe, Caucasia, the Middle East and South West Asia have all but vanished. Today, just a tiny number of them survive in one single country outside of Africa.

That country is India, a land where lions were once widely distributed and considered a common sight. However, by the late 1880s, they had become more conspicuous by their absence, prompting the British naturalist and geologist William Blandford to note in 1891 that in India 'the lion is verging on extinction'.

India's lions were lost first from the north in the 1830s, when the last lion was seen in Delhi in 1834. By the 1840s, they were gone from the east too. In central and western India, they survived a little longer, disappearing from central areas, including the Vindhya Range and Bundelkhand regions, in the 1860s.

At Rajasthan in the west, just a few lions were thought to survive into the 1870s. A decade later they were gone from the western Aravalis, leaving India's only remaining lions in the south. From here most of them soon vanished too. By the 1900s, only around a dozen Asiatic lions, located in the Gir Forest at Gujarat in the Saurashtra Peninsula in the southwest of the country, remained on earth, facing imminent extinction.

Fortunately, since then various protection programmes have brought the Asiatic lion back from the brink of extinction. Today a small population of around 520 Asiatic lions remain at Gir and the surrounding areas. In 2002, lions in India were listed on the IUCN Red List as critically endangered. Eight years later, in 2008, as numbers continued to grow at Gir, the lions were downgraded and listed as endangered.

Even though lions are well protected at Gir, their total population size is small and they remain vulnerable to the consequences of inbreeding, as well as losing large numbers in the event of disease or natural disasters such as forest fires and flooding.

The millennia-long dispersal of the world's lions has had a sad end. Theirs isn't a journey to other countries or other continents; theirs is a dispersal to extinction, to nothingness. No other species on earth has lost so much of its ground, seen its former territories shrink over historical times, as the lion has.

Lions in India are endangered. Lions in West Africa are critically endangered and those in East and Central Africa are considered regionally endangered. Only lions in the south of the continent are considered safe.

The numbers are stark and they are unforgiving. Just over 500 lions remain in India. Fewer than 20,000 survive in Africa, its once indomitable heartland. All live in tiny splinters of their former ranges. Lion numbers have plummeted and continue to fall.

The lions we have left are remnants, just scraps of lions found here and there, bits and pieces of once vibrant lion populations, a ragged tapestry, once rich and golden, now fading before our eyes. The bright flame of an iconic species is burning out.

CHAPTER THREE

People Hate Lions – Part I

I am Ashurbanipal, the King of the World, the King of Assyria! For my regal amusement I have caught the lion by the tail, and on the instructions of my helpers, the Gods Ninib and Nergal, I have split his head with the two-handed sword.

King Ashurbanipal of Assyria, 685–627 BC

We now live in a world with precious few lions. Both the lions and the endless stretches of grassy plains and savannah they roamed in their hundreds of thousands, less than a century ago, are gone.

An undeclared war, millennia old, has left us with fewer than 20,000 lions. It's a war that has taken many guises. In their thousands, lions have been removed from the wild, imprisoned and publicly executed. They have been massacred, slaughtered and poisoned. They are still hunted down and beheaded for display or for punishment.

As a result, Cape and Barbary lions have been eradicated. Our last lions are on the edge of extinction. The only lion extermination that humans aren't conclusively implicated in is the dying out of the Ice Age lions.

In Chapter Two, we saw how the Late Quaternary Megafaunal Extinctions (LQME) project, headed by Tony Stuart and Adrian Lister, provided terminal dates for the lions of the Ice Age. As well as generating an extinction timeline, the project also considered the factors that may have caused the disappearance of the Ice Age's big cats from Eurasia and North America.

Their conclusions are shared in their research paper 'Extinction Chronology of the Cave Lion' (also referred to in Chapter Two). Although the reasons for the loss of the species can't be pinpointed precisely, say Stuart and Lister, there is strong evidence to suggest that climate change could have contributed to their demise.

Around 15,000 years ago, when the big chill of the Ice Age gave way to warmer temperatures, the great ice sheets of the northern hemisphere began to thaw and retreat, flooding Europe, Asia and the Americas, altering their landscape. Grassy plains and open steppes filled up with shrubs and trees growing closely together, forming dense forests.

Lions, used to bringing down prey in an uncluttered landscape, began to struggle in these more confined environments. When the arenas where lions once triumphed began to close up, they now struggled to catch prey. As their larders went bare, lions went hungry, wasting away, slipping into eventual extinction, most probably as a result of starvation.

While Stuart and Lister can't rule out other factors contributing to the extinction of cave lions, they do point to a definite correlation between the warming up of the climate and 'a spread of shrubs and trees and reduction in open habitats' that quite possibly led to the end of the line for the species.

Previously, the fall of cave lions had often been attributed to the extinction of their prey. However, as Stuart and Lister point out, this isn't likely. Although a significant number of creatures as well as lions did disappear by the close of the Ice Age, much of their prey, including horses, musk oxen and red deer, survived for thousands of years after them. Some, like reindeers and bison, are still with us today. The fact that so many of the lions' prey survived should rule this out as a theory now, state Stuart and Lister.

As their world defrosted and their ranges diminished, lions not only found hunting more challenging, they also found it harder to avoid the spears of early humans. Although we can't be sure if early humans were killing lions in self-defence, for food or for other reasons, we do know that hunting in some form was happening.

In south-west France, deep inside the Lascaux Cave, there's a section of wall engraved with images around 16,000 years old of cave lions. Two of these lions appear to have spears wedged in their sides. One of them looks like it is coughing up blood in a last dying throe. To the left of the speared lions there's a third lion with a cross marked on it. Possibly, some archaeologists have suggested, the scene depicts a lion hunt and the big cat marked with the cross was a lion that got away from its human hunters.

Around 10,000 years earlier, at Dolní Věstonice, an archaeological site in Moravia in the east of the Czech Republic, other Ice Age artists were at work. There, one of the artists, using clay mixed with bone, sculpted a tiny figurine of a lion. Today, only the lion's petite ceramic head, measuring just under 3 centimetres, survives. It was found alongside hundreds of other animal figurines by archaeologists excavating the open-air site in the mid-1920s.

What they found especially interesting about the little disembodied head, apart from the exquisite detail of the lion's face, were two holes: one by an eye and the other above one of the lion's ears. The holes would have been deliberately made while the figurine was still wet, just before it was dried and fired in a kiln. They're believed to represent wounds caused by spears, another suggestion that early humans may have been hunting cave lions.

In 2000 and 2001, at the Gran Dolina Cave site in the Sierra de Atapuerca region of central Spain, Ruth Blasco (an archaeologist then based at the Universitat Rovira i Virgili in Tarragona, north-east Spain) and her team unearthed seventeen bones of an ancestral cave lion they believe to have been hunted, killed and eaten by early humans.

In her 2010 research paper 'The Hunted Hunter: The Capture of a Lion at the Gran Dolina Site, Sierra de Atapuerca, Spain', Blasco reports that the scars on the bones are cut marks made by humans when the lion was skinned and

defleshed before being eaten. She also points out that the way
the lion had been gutted and its bones broken to get at the
marrow was identical to the way that early humans processed
bison, deer and horses, other animals known to be eaten by
early humans.

After her study was published, Blasco spoke with *National
Geographic*, explaining that this could mean that '*H. heidelbergensis*
[early human] was a top predator of its day, and could hunt and
even kill the deadly cave lion'.

Not all scientists are convinced, though. Some are sceptical
about early humans' hunting prowess and their ability to
overcome fearsome Ice Age lions. Perhaps the Gran Dolina lion
died of natural causes, had an accident or was killed by another
animal (quite possibly another lion), it has been suggested.

'Lions die of all kinds of causes, and one can't rule out early
access scavenging of a natural death,' John Shea, a
palaeoanthropologist at Stony Brook University in New York
in the United States, told *National Geographic*. In response,
Blasco pointed out that the lion bones they found showed no
signs of illness or injury, and that if 'other animals had killed
the lion, the tasty viscera would have been long gone by the
time the early humans arrived'.

Daniel Adler, an anthropologist from the University of
Connecticut in the United States, was also not convinced,
remarking to *National Geographic* that even if early humans at
Gran Dolina 'hunted this cave lion, no other examples exist
elsewhere in Eurasia'.

However, archaeologists have unearthed Ice Age lion
bones in the camps of early humans in North America. At
Jaguar Cave in the Beaverhead Mountains of Idaho in the
north-west of the United States, American lion bones have
been found in the sediments from areas once used by Palaeo-
Indians for dumping their food refuse, suggesting lions may
have been on their menu, too.

In 2016, new evidence that people may have been hunting
cave lions was revealed when a research paper called 'Under
the Skin of a Lion: Unique Evidence of Upper Paleolithic

Exploitation and Use of Cave Lion from the Lower Gallery of La Garma (Spain)' was published in *PLOS ONE*.

The study, led by Marián Cueto, a zooarchaeologist at the University of Cantabria in northern Spain, detailed the find of 9 fossilised claws belonging to a single cave lion, dated as being around 16,000 years old, at the La Garma Cave complex in Cantabria.

The focus of the study was on the cuts and scrapes found on the claws. Close examination showed they were made by stone tools, not dissimilar to the marks made by hunters today when they skin their trophies. The marks were also precise and consistent, suggesting that the woman or man who skinned the lion was highly skilled and familiar with lion anatomy. Commenting on the find, Edgard Camarós, co-author of the study, told *The Guardian* that because claws had been left attached to the skin it was highly likely that the skin was used as a pelt, perhaps as a rug (the skin was found on a cave floor) or even as a roof for a simple hut.

It's also possible the skin was used in ritual activity. Talking with *Smithsonian* about the likelihood that the skin could have been used ritually, Cueto told the magazine that because lions were challenging and dangerous beasts to hunt 'it probably played an important role as a trophy and for use in rituals', while also noting that many early people have 'used carnivore pelts as a symbol of power'.

However, other archaeologists remain unconvinced. Hervé Bocherens, a palaeobiologist at the University of Tübingen in south-west Germany, told the same magazine that while the lion's skin may have been used as part of a ritual, more evidence is needed to show that this practice was widespread among early humans. For example, Bocherens wanted to know if the lion was local or had been 'obtained from other prehistoric groups'. Without more data, Bocherens maintained that no conclusions could be drawn about the use of the skin.

A similar concern was voiced by Clive Finlayson, palaeoanthropologist and director of the Gibraltar Museum in southern Spain, who told *The Guardian* that without more evidence, the killing of this lion 'may have been a one-off. For now, we simply don't know.'

Although evidence does suggest that Ice Age people were killing cave lions, possibly for food or for their skins, it's unlikely that hunting alone would have caused the species to become extinct. Rather, it's a culmination of correlative events. With the climate warming and forests replacing grassy plains, lions lost the killing fields over which they had evolved to rule.

As the lions became fewer in number and weakened by hunger, early humans took advantage of the once mighty cave lions' new vulnerability, speeding up – rather than causing – the lion's exit from the world, while they themselves started their ascent towards top predator status.

While early humans may not have been solely responsible for the end of the Ice Age lions, the relationship between modern humans and later lions, *Panthera leo*, is bloodstained. Our fingerprints are all over their diminishment.

Since ancient times, humans have hunted lions. As well as being hunted by pastoralists and farmers to protect their livestock, they've also been hunted for spectacle, for vanity and for entertainment. For rulers of ancient kingdoms – be they pharaohs, emperors or kings – hunting lions was the best way to display their god-given uniqueness, their superiority and distinctness from ordinary people. If a king could hunt down and kill a beast as fabulous and mighty as a fearsome lion, if he could end the life of the king of the beasts, then he himself as king of kings surpassed nature and possessed an innate and indisputable right to rule.

The earliest records we have of lion hunting as a royal pursuit come from ancient Egypt, from the reign of Amenhotep III, who ruled from 1386 to 1349 BC.[*] They

---

[*] The dates of Amenhotep III's reign vary. His reign is also cited as being from 1388 to 1351/1350 BC.

take the form of stone plaques, shaped like scarab beetles and known as scarabs, with text carved into them. To keep his people informed about his latest achievements, Amenhotep commissioned hundreds of scarabs inscribed with his royal news. Around 200 of these 'palace bulletins' have been found in Amenhotep's former kingdom, some as far afield as Syria to the north and Sudan to the south.

Of the 200 scarabs, 123 record Amenhotep's lion hunt tally, making them the world's oldest record of a big-game hunter's prowess. The text on the scarabs tells us that during his first 10 years as ruler, Amenhotep brought down '102 fierce lions ... with his own arrows', an impressive feat that Amenhotep was keen to share with his subjects, adding to his stature and reinforcing his reputation as a great and mighty king.

Regarding the lion-hunting exploits of other pharaohs, we know very little. In *Lion*, the historian Deirdre Jackson tells us that Amenhotep's 'use of scarabs to record his hunting successes is unparalleled. No other Egyptian pharaoh did likewise.'

The theologian and historian George Rawlinson writes in *The Story of Ancient Egypt*, published in 1897, that the only other pharaoh who makes a boast of his hunting prowess is Amenemhat I, who reigned from 1991 to 1962 BC,* some 600 years before Amenhotep III came to the throne. About Amenhotep III's lion-hunting, Rawlinson tells us: 'When Amenhotep was not engaged in hunting men his favourite recreation was to indulge in the chase of the lion ... Later on in his reign he presented to the priests who had the charge of the ancient temple of Karnak a number of live lions, which he had probably caught in traps.'

Rawlinson also adds that a good many of Amenhotep's hunts probably took place in Mesopotamia, where lions

---

* Dates for Amenemhat I's reign vary. For example, some historians cite 1985 to 1956 BC while others cite 1994 to 1964 BC as the years he ruled.

abounded and where he went frequently to capture men and
women, bringing them back to Egypt as slaves.* His hunts
would have been well executed, military-like affairs, notes C.
A. W. Guggisberg in *Simba: The Life of the Lion*, published in
1961. An authority on lion history and behaviour, Guggisberg
tells how pharaohs shot lions with bows and arrows from war
chariots, accompanied by bodyguards, gamekeepers and 'a
whole army of nobles'.

Lions were also royal sport for the kings of ancient Assyria,
whose vast kingdoms covered much of the Middle East† as
well as Cyprus, Turkey, and Egypt and Sudan in North
Africa. Assyrian kings, particularly those 'who ruled Assyria
from the ninth to seventh centuries BC were great lion
hunters', says Guggisberg in *Simba*. Like the pharaohs, they
hunted from chariots accompanied by an entourage of
'horsemen, foot soldiers, and kennel men [and] their mastiff-
like hunting dogs'.

One of these great lion hunters was King Ashurnasirpal II,
who ruled from 883 to 859 BC. Keen that his people should
know about his great skill at hunting lions, Ashurnasirpal II
commissioned portraits of himself carved into alabaster reliefs
to decorate his palace walls.

One of these reliefs, now displayed at the British Museum
in London, shows Ashurnasirpal II in his chariot taking aim
at a lion at close quarters with a bow and arrow. Alongside
the image, text details the king's personal hunting tally for
865 BC, including 370 'great lions … killed with hunting
spears', 257 wild oxen and 30 elephants.

Reliefs like this helped reinforce the king's intrinsic
majesty, reminding his subjects that he was king for good
reason. An even better way for a king to assert his divinity

---

* Today this corresponds approximately to regions along the Iran–
Iraq and Turkish–Syrian borders, most of Iraq as well as Kuwait,
eastern Syria and south-eastern Turkey.
† Iraq, Syria, Saudi Arabia, Jordan, Iran, Kuwait, Lebanon and Israel.

was to hunt lions publically, so people could see for themselves that the right man was indeed on the throne.

The only risk with this strategy, of course, was not bagging any lions. To ensure this didn't happen, a later royal, King Ashurbanipal, who reigned from 668 to 627 BC, took no chances when it came to making lion hunting a spectator sport. We know this because Ashurbanipal also commissioned reliefs depicting great events from his life, including a spectacular public lion hunt.

Collectively known as the Lion Hunt of Ashurbanipal reliefs, around 48 carved scenes bring to life from start to finish a dramatic royal hunt that took place almost 2,500 years ago. They were unearthed in the early 1850s by Hormuzd Rassam, an Iraqi archaeologist, at Ashurbanipal's palace in Nineveh on the outskirts of Mosul in modern-day northern Iraq.

The reliefs, created between 645 and 635 BC, were taken to the British Museum, where they are still exhibited. It's at the British Museum that I'm able to appreciate why they're considered masterpieces of Assyrian art and learn for myself just how Ashurbanipal was able to slay numerous lions and grab guaranteed glory for himself.

British Museum, London, England
King Ashurbanipal's reliefs hang in panels along the sides of a narrow gallery, its ceiling dotted with bright spotlights that pick out the detail of the scenes in each panel. As I walk past each one, the story of the hunt unfolds before me, like the pages of a picture book.

From the start of the chase, I can see evidence that the hunt is carefully controlled. Large crowds are shown on a hill looking down into an arena enclosed with barriers. In the centre of the arena is Ashurbanipal. He's sitting astride a handsome horse, wearing his finest attire, looking noble and invincible.

Surrounding Ashurbanipal, in chariots and on horseback, are his huntsmen. Lined up in a semi-circle, armed with bows and arrows, spears and swords, they face a line of riled-looking lions, caged in wooden crates. Many of these lions would have been bred in captivity specifically for hunting, the offspring of lions taken from the many lands of Ashurbanipal's lion-rich empire – canned hunting, the ancient way.

As the lions are released, their bolt for freedom is met with a barrage of arrows, tips dripping with deadly poison. Lions caught cowering by their cages are prodded and whipped until aggression replaces fear, and they too join the charge. As the lions are hit by arrows and spears, they stagger and fall. One vomits blood. Another, a lioness, her spinal cord severed by an arrow, struggles to move as paralysis shuts down her body. Others manage to drag themselves across the ground on their stomachs, their legs useless and spent.

Lions that manage to dodge the lethal arrows aren't able to escape the arena. They meet a wall of soldiers with spears and snarling mastiffs and are forced back to the slaughter. As the lions grow visibly weaker, Ashurbanipal dismounts from his horse and wields his sword. Then, one after another, any lion found still breathing has its throat cut, courtesy of the royal sword.

The final panels of the reliefs show Ashurbanipal with the bodies of four large male lions. He is standing at an altar, pouring wine over the four corpses to honour Ishtar, the goddess of war and to whom the lion was sacred. Caption-like text inscribed by the scene reads: 'I am Ashurbanipal, to whom Assur and Belit have given sublime powers. On the lions I have killed I have set the grim bow of Ishtar, mistress of battle, I have offered a sacrifice and poured wine over them.' Similar allusions to Ashurbanipal's god-like status are made in other inscriptions too.

Throughout the panels, Ashurbanipal is always taller than the huntsmen and soldiers who surround him, rising higher than other mortals, both physically and spiritually. His lion slaying is a symbol of his vigour and virility. As a conqueror of nature, Ashurbanipal is all but deified. An element noted by Jackson in *Lion* when she states: 'One of the primary purposes of the Assyrian lion hunt was to acknowledge the supernatural power of the king.'

One of the most moving aspects of the reliefs is the way the lions and horses are depicted. The fear felt by both species is tangible. Horses roll their eyes upwards as if trying to block out the sight of the massacre in front of them. They rear up on their hind legs, muscles tensed tightly as their handlers try to reassure them. Their reluctance to move forwards suggests they're also bothered by bewildering noise, the raucous crowd, shouting and cheering, the chariots thundering by and soldiers huffing and puffing. And, of course, the sound of lions growling and roaring, running for their lives and breathing their last.

The agony of the lions feels as real as the horses' fear. Their faces twist in pain, their bodies racked with exhaustion. They buckle and fall. Claws scrape at the ground and jaws open wide, bone cracking and splintering as they let out their last roars.

Ashurbanipal, in contrast, remains stoic. His face never shows emotion. He stays ever tall, his backbone straight and unfaltering like a statue. Remaining strong in the face of the pain and the terror that surrounds him, contrasting with the weakness of the natural world around him.

Despite his steely aspect, Ashurbanipal is shown to have a beneficent side. Aware that his subjects and their livestock are often attacked by lions, in text accompanying one of the panels, Ashurbanipal lets his people know that he understands their plight and has taken action: 'Lions ... have bred in mighty numbers ... They constantly kill the livestock of the fields, and they spill the blood of men and cattle. The misdeeds of these lions have been reported to me ... [so] I have penetrated their hiding places and destroyed their lairs.' Ashurbanipal makes it clear that as well as being a supernatural slayer of lions, he also has the welfare of his people at heart, protecting both them and their livelihoods from the jaws of such a beast. This, combined with the spectacle of his public lion hunt – albeit a carefully choreographed bloodbath, a pageant of propaganda guaranteeing Ashurbanipal's reputation and glory – confirmed to the Assyrian people that he was the right man to be wearing the crown. Only a true king could conquer the king of the beasts.

The Barbary lion's decline in North Africa began around 264 BC when the Romans started travelling to the continent in search of exotic wild animals, including lions, leopards, elephants, rhinos, hippopotamuses and crocodiles, to capture and display back home in the capital.

Catching Africa's wild lions was a job for the skilled and certainly not for the fainthearted. There were no tranquillisers and beasts could not be speared, as they needed to be in prime condition to impress the people of Rome. The best ways to capture lions was either by trapping them in pits or catching them in nets.

In his work *The Chase*, Oppian, a second-century BC Greco-Roman poet, writes about the excellent lion hunts in Libya and elsewhere in North Africa. He describes the Roman lion hunters as greatly daring men with valiant spirits, and he details how lions were trapped in pits. First, Oppian says, a place was found where a 'well-maned roaring Lion dwells'. Then a wide and large round pit was dug out and a wall of close-set boulders built around it so that 'the Lion may not see the crafty chasm'. Inside the pit a great pillar, sheer and high was then built, from which a suckling lamb taken from its mother was hung aloft.

The lamb's forlorn bleats would then strike the 'Lion's hungry heart' and, obeying the impulse of hunger, the lion would leap over the wall, finding itself in 'the gulf of a pit unlooked for'. At this moment, hunters hidden to the lion would appear and 'with well-cut straps ... let down a plaited well-compacted cage, in which also they put a piece of roasted meat'. Unable to resist the smell of the meat, the lion would then leap into the cage exulting, only to find itself trapped.

When nets were used to capture lions, men on horseback carried blazing torches and banged drums to steer lions towards hunters standing by with staked nets. When a lion approached them, the hunters would throw their net over the big cat, securing it to the ground with stakes, then bundling the netted lion into a waiting cage to be transported back to Rome.

These methods of capturing lions, and other wild animals, proved so successful that American author Robert M. McClung in his book *Lost Wild Worlds* notes that by the reign of Augustus (63 BC to 14 AD), lions had become so scarce that Libyan agriculturalists were able to tend their crops with 'comparatively little fear of attack from the big cats'.

The Romans' demand for lions and other exotic beasts created an empire-wide industry. Hunters, captors, shippers, keepers, handlers and trainers were all recruited and trained to snatch thousands of lions from the wilds of North Africa and confine them in cages and dungeons all over the Roman Empire. The naturalist David Attenborough writes in his book *The First Eden: The Mediterranean World and Man* that the demand for lions and their subsequent capture was so extensive 'it may well have been a significant factor' in the disappearance of lions from Roman strongholds in North Africa.

In the early days of bringing lions back from North Africa, wild animals were initially displayed at the public games put on for the people of Rome, either by the state or by wealthy and influential individuals. The games, which often lasted for days at a time, sometimes weeks, took place around Rome at purpose-built stadiums and arenas, the most famous including the Colosseum, which could house up to 80,000 spectators, and the huge Circus Maximus that could hold 150,000.

The programmes of the games were varied but always spectacular, typically featuring chariot racing, animal parades and hunts, gladiator fights and the re-enactment of sea battles, as well as public executions. Originally, animal parades were used to honour the dead and included birds and beasts indigenous to the Italian peninsula. However, when exotic animals like lions, elephants and leopards began to arrive in Rome, native animals lost their appeal. Far more fascinating were the weird and wonderful animals arriving from Africa, the likes of which had never been seen before.

Although initially thrilled to see such fabulous beasts, the crowds soon wanted to see something more exciting than just animals being displayed or paraded about. Over time such animals lost their exotic appeal. People demanded something more exciting, something more dramatic to whet their theatrical appetites.

It was the crowd-pleasing Marcus Fulvius Nobilior, a Roman general, who gave the audience what it wanted. In 186 BC, Nobilior added a new event to the public games he was putting on to celebrate a recent military victory. His *pièce de résistance* was adding to his line-up the new experience of watching gladiators fighting lions and panthers.

The addition was so remarkable that it was included in *The History of Rome*, an epic account of Roman history written at the turn of the first millennium by the historian Titus Livius, commonly known as Livy. Of Nobilior's games, Livy wrote: 'For ten days, with great magnificence, Marcus Fulvius gave the games … a contest of athletes was then for the first time made a spectacle for the Romans and a hunt of lions and panthers was given.'

Such a spectacle delighted the crowds, and animal hunts – or *venationes*,* as they became known – soon became a staple part of public games. *Venatores*,† the gladiators who took on lions and other wild animals, were highly skilled in animal combat, having learned the tricks of the trade at a specialist gladiator training school called the *Ludus Matutinus* or

* Latin for 'animal hunts'.
† Latin for 'hunter'.

morning school; so called because *venationes* were always staged in the mornings, followed by top-of-the-bill gladiator battles in the afternoon.

Venatores were trained to use a number of weapons, including bows and arrows, tridents, swords and daggers, in addition to a variety of wooden hunting spears, including a 2-metre-long spear known as a *venabulum*, and a *pilum*, a heavy javelin of approximately the same length. With these fighting tools, *venatores* were able to slay all kinds of animals, from shy gazelles to snarling lions, as well as snapping crocodiles and enraged elephants.

When taking on a large number of lions at the Colosseum, *venatores*, kitted out with helmets and shields, would line up in flanks, just like an army, and wait for the lions to appear in cages from below ground. In *The First Eden*, Attenborough describes how lions were lifted from their dungeons up into the centre of the arena in 'ingenious' iron cages, of which the Colosseum had thirty-two.

As the big cats were released, drums were beaten, and when the lions roared at the audience, the crowd roared back. What the spectators didn't know was that below ground, the lions had been deliberately riled. They were prodded at with poles and often starved prior to fights, making them both angry and hungry.

Around shrubs planted to create a feel of the wild, lions darted and dodged, trying to avoid the carefully targeted aim of spears thrown by adrenalin-fuelled *venatores*. By the end of the hunt, the arena would have been strewn with dead lions, their manes matted with blood from dagger wounds to the head, their sides gashed open with spears and hearts brought out on sword tips.

Once the *venationes* were over, surviving lions would also make an appearance in the *damnatio ad bestias*,[*] an especially bloodthirsty lunchtime event that took place before the gladiatorial fights. These were public executions where

---

[*] Latin for 'damnation to beasts'.

condemned criminals, prisoners of war and some early Christians, unarmed and naked, were thrown to ravenous lions and other wild animals.

Known as *bestiarii*,* these men and women were quickly killed and eaten in front of a cheering, unforgiving crowd. Their plight was recorded by Apuleius, a writer and philosopher, in his novel *The Golden Ass*, written during the time of the early Roman Empire: 'Criminals destined for a fate without hope were nevertheless well fed in order to fatten the animals ... A special effort had been made to bring these brave animals from abroad to serve as executioners for those condemned to death.'

Putting on fabulous games was essential. They were a public way of showcasing power, leadership, influence and wealth. The more extraordinary an event was, the more extraordinary the man behind it was assumed to be. And one way of making games remarkable was by putting on an extraordinary *venatione*.

When a fantastic creature like the Barbary lion, with its full dark mane and muscular torso, joined the cast of animals taking part in *venationes*, people flocked to see the new feline star of the hunting show take on Rome's finest *venatores*.

Lions were the A-list animals of hunting-show casts. Not only were they terrific crowd pleasers, but they also demonstrated the reach and power of both the Roman Republic and Roman Empire; not just over the human world, but the natural world, too. When lions appeared in arenas, audiences, tens of thousands strong, stood up and cheered. And when they cheered for the lions, they were cheering for their leaders, too.

Emperors, statesmen and generals were acutely aware of this. For some of them, lions – lots of them – were essential for creating outstanding *venationes*. For example, in 55 BC, Pompey's *venatores* slaughtered 600 lions in one single show. The feat was noted by the Roman historian Pliny the Elder in

* Latin for 'beast fighters'.

Book VIII of his encyclopaedic *Natural History*, in which he wrote: 'Pompey the Great had 600 lions in the Circus, 315 of which had manes.'

In *Animals for Show and Pleasure in Ancient Rome*, published in 2005, the historian George Jennison comments that for Pompey to maintain a god-like reputation, it was essential that he provide such spectacles for the people of Rome. So, in 55 BC, Pompey's games featured an 'unprecedented show of animals', including hundreds of lions and leopards, large numbers of elephants, and animals that Romans had never seen before.

In a similar vein, Jennison says Julius Caesar, too, 'had to entertain the Roman public in ways that would express that he was Caesar and unique'. Which is why Caesar's *venationes*, 11 years later, included a hunt of 400 lions, as well as fights with elephants and bulls and the parading of the first giraffe ever seen in Rome.

It wasn't until the fifth century AD that the staging of public games began to wane. The last great gladiatorial duel took place at the Colosseum in 404 AD. The *venationes*, although banned from much of the Empire in 498 AD, were one of the last Roman spectacles to go. Rome's last *venatione* was in 523 AD, and the final one of the Empire was held at Constantinople in the south fourteen years later.

Despite the ban, *venationes* continued, but in a different format than before. Now their savagery was diluted. Instead of the focus being on the death of an animal by a *venatore*, the emphasis was, says D. L. Bomgardner, author of *The Story of the Roman Amphitheatre*, 'not on the death of the animal, but successful escape from its teeth and claws by the performers involved'.

The 'gentlification' and final demise of the *venationes* was, like the end of the gladiatorial games, a sign of the times. Under the growing influence of the Christian Church, which disapproved of such bloodthirsty events, social attitudes changed. At the same time, the Roman Empire, which had for centuries been under immense strain having faced invasion, civil war and plague, leading to severe economic crises, could no longer afford to put on such shows.

Venationes were incredibly expensive. As wild animals started to become scarce, capturing them became more difficult and increasingly costly. It's likely, as Bomgardner suggests, that the *venationes* finally came to an end as a result of both social and financial pressures.

In India, during the Mughal Empire (1526 to 1540 and 1555 to 1857 AD), hunting lions was strictly the reserve of emperors. Ruling over a 'fabulously rich empire', their hunts, known as *shikar*,* were extravagant theatrical affairs, described by Guggisberg as being 'of a splendour and scale rivalling those held at the time of the Pharaohs and Assyrian Kings'.

While the pharaohs had the stories of their lion chases inscribed into stone scarabs and Assyrian kings had their battles with lions carved in stone reliefs, Mughal emperors had their encounters brought to life in elaborate paintings, exquisite book illustrations and vividly written memoirs and court journals. Many of these are still with us and provide fascinating insights into the Mughals' pursuit of lions.

Emperors and their exotic and colourful posse travelled around India in organised *shikar* that could last weeks, sometimes even months. Just like the hunts of the Assyrian kings, emperors were always set up to win. Ahead of the hunt, fences were put up around known lion haunts to make escape impossible as well as to ensure there were enough lions around for the emperor to take aim at.

The Emperor Shah Jahan, who ruled from 1628 to 1658 AD, commissioned a royal book called the *Padshahnama*† to detail his life's great achievements. The book contains more

---

* Urdu for 'hunt'.
† The *Padshahnama* is held by the British Royal Collection at Windsor Castle in the south-east of England.

than forty pictures painted by fourteen court artists between 1630 and 1657 AD. Described as some of the finest Mughal paintings ever produced, the book's illustrations chronicle Shah Jahan's reign.

Some of the plates in the book show details of lion hunts. One of them, called *Shah Jahan Hunting Lions Near Burhanpur*, shows the emperor, astride an elephant, taking aim at a lion standing near a lioness and two cubs. Part of the area is fenced off by a *bëdar*, a very strong net secured to the ground by large poles, effectively cornering the small pride of lions, making it easy for Shah Jahan to bring the big cats down.

There's another depiction of a lion hunt in a wall painting[*] that also features Shah Jahan, this time accompanied by his sons. The area the royal family is hunting in is enclosed by a *bëdar* and, in addition, sizeable buffaloes are stationed around the edges to further prevent the lions escaping.

Another wall painting called *Aurangzeb Hunting Lions*, held at the Chester Beatty Library in Dublin, Ireland, shows Emperor Aurangzeb,[†] who ruled from 1658 to 1707 AD, taking aim at a lion, also in an area sealed off with nets. Aurangzeb's other hunting methods are recorded in *Travels in the Moghul Empire, A.D. 1656–1668* by François Bernier, a French doctor who travelled in India and worked as royal physician for the emperor.

In the travelogue, Bernier describes how lions were drugged to make them easier prey for Aurangzeb. In advance of royal *shikar*, Bernier says the emperor's gamekeepers would tether a donkey to a tree in a spot known to be visited frequently by a lion, which in due course would be eaten by the visiting big cat.

Then every day for the next few days another donkey was tied to the same tree and was again eaten by the returning lion, which no longer needed to seek out other prey. When

---

[*] The painting is unnamed, as is its painter, and is held in a private collection.

[†] The son of Emperor Shah Jahan.

learning that Aurangzeb's visit was imminent, the gamekeepers would then put out a donkey 'down whose throat a large quantity of opium ha[d] been forced'.

Once the intoxicated donkey had been eaten, it 'produce[d] a soporific effect upon the lion', enabling the gamekeepers to throw a net over it and confine it to one spot until Aurangzeb and his entourage, armed with half-pikes, arrived. From his elephant, the emperor would then take shots at the netted lion until it was 'at length killed'.

One of India's most prolific Mughal hunters was the Emperor Jahangir, who ruled from 1605 until 1627 AD. During a hunting career that lasted almost 40 years, he killed more than 17,000 animals, including 86 lions. Compared to the rest of his wild-beast tally, the number of lions Jahangir killed seems relatively small. But despite the relatively paltry number, Jahangir was incredibly proud of his lion bounty. In his diary, he kept a list meticulously detailing all the animals he had slain. The eighty-six lions were placed at the top of that list, reflecting their importance to Jahangir.

Under the Mughals, lions had been royal quarry for more than 300 years, but from the late 1600s when the British began to arrive in India, the threat from hunting grew in magnitude and brought lions in India to the brink of extinction.

Quite simply, as Guggisberg points out in *Simba*, when the British came to India 'they enthusiastically took over where the Mughal emperors left off'. The fact is reiterated by Divyabhanusinh, author of numerous books about Asiatic lions, in *The Great Mughals Go Hunting Lions*, where he notes that the lions 'left behind by the Mughals [were soon] decimated by their successors', the British.

The big decline in lion populations in India started around 1757 when the East India Company established 'company rule' and started to use its own private armies to rule large parts of India. With the rise of the British Empire, lion hunting was no longer reserved for royalty. East India Company employees – and members of the British Army – were encouraged to indulge in *shikar* to build character and fortitude, instead of taking up more harmful activities such as gambling and opium use.

In *Ecology and Biogeography in India*, biogeographer M. S. Bani describes how wildlife began to suffer as the British started trophy hunting, bagging lions, tigers, leopards, rhinos and elephants without restraint.

Although as Jackson notes in *Lion*, 'accounts of lion hunting in India in the modern period are in short supply', we do have records of some colonial hunters' lion counts. For example, in *The Story of Asia's Lions*, Divyabhanusinh shares some known hunting scores from the 1800s that include details of Colonel James Skinner hunting lions from his horse, of Andrew Fraser who ended the lives of 84 lions, and a George Acland Smith who killed a record-breaking 300 lions near Delhi in 1857.

Guggisberg also details reports of a prolific lion hunter called Sir Bartle Frere, an English colonial administrator who went on to become Governor of Bombay in 1862, but who in the 1840s, with the help of his brother, shot 'seven or eight Indian lions ... in the course of a single day'.

In 1878, wildlife was threatened further when the British introduced the Indian Forest Act, which turned around a fifth of South Asia into hunting ground, providing a seemingly endless supply of timber and animal products, including horns, bones and skins, to ship back to Blighty.

With licence to hunt what and where they liked, colonialists simply shot lions until there were no more left to shoot. By the early 1900s, the only lions left in Asia could be found at the Gir Forest in the Kathiawar region of western India, where in 1901 Lord Curzon, the Viceroy of India, was invited to join a lion hunt.

Although a keen hunter, Curzon was aware of and concerned about the plummeting lion population and declined the invitation politely. The following year, in 1902, he wrote to the Bombay Game Preservation Society to raise awareness of their plight, noting that lions, once widespread, were 'now confined to an ever-narrowing patch of forest in Kathiawar'. Curzon was right to be concerned. Asiatic lions were on the brink of extinction, their future as a subspecies hanging in the balance as a direct result of colonialists and their guns.

While the hunts of the Egyptians and the Romans contributed to the decline of lion populations in Egypt and Libya in North Africa, Barbary lions in Morocco, Algeria and Tunisia were left relatively unscathed. It wasn't until the seventeenth century AD that the lions in these countries came under threat.

The Spanish traveller Luis del Mármol Carvajal, for example, journeying through Morocco in the 1500s wrote in his book *General Description of Africa*: 'There are so many lions ... that they are not feared.' A century or so later, the Scottish cartographer John Ogilby wrote in his book about North Africa (published in 1670) that in Tunisia, the 'whole country, especially in its mountainous parts, is full of lions'. And in *Simba*, Guggisberg notes that in Algeria during the early 1800s lions were still 'very numerous'.

However, as the twentieth century approached this was no longer the case. Count Saint-Marie, a French military man travelling in Algeria in the late 1800s, noted in his book *Algeria in 1845: A Visit to the French Possessions in Africa*, that lions in the country 'are not very common'. In 1881, lions in Tunisia remained in the farthest north-west, and in Morocco the country's last lions had 'retreated everywhere from the coastal area' and were gone from the north by the 1880s.

This change in the fortune of lions in Morocco, Tunisia and Algeria can largely be attributed to the introduction of hunting bounties from foreign governments. Under Turkish rule, hunters in both Tunisia (1574 to 1881 AD) and Algeria (1550 to 1830 AD) were financially well rewarded for bringing down big cats. Lions, problem killers of livestock, were considered an economic liability and a threat to national prosperity that needed to be removed swiftly.

Writing about the decline of lions in North Africa in *The Book of the Lion*, published in 1913, Sir Alfred Pease, a British politician and well-known lion and big-game hunter of the time, stated that lions in the region were 'greatly diminished' largely because of the 'high price placed on their heads by the Turkish Government in the Barbary States'.

The Turkish government also gave Algeria's two great lion-hunting tribes much encouragement to hunt more lions, exempting them from all taxes and providing them with substantial financial rewards for their skins.

When the French took over from the Turkish – in Algeria in 1830 and Tunisia in 1880 – the bounties continued. They were then introduced in Morocco in 1881, when French rule later started there. From Algeria, there are government records that show the decline in the number of the country's lions being submitted for bounties. In *Great and Small Game of Africa*, published in 1899, the hunter and naturalist Henry Anderson Bryden provides the exact numbers: 'Between 1873 and 1883 the process of extinction is measured in Government returns. The numbers killed for the whole of Algeria were, in the six years of this time period, 1878, 28; 1879, 22; 1880, 16; 1881, 6; 1882, 4; 1883, 3 and 1884, 1.'

Hunters from France and elsewhere in Europe came to the region specifically to hunt lions. Jules Gérard, famous in France for his lion-hunting pursuits, spent time in Algeria in the 1800s tracking and killing the big cats. While in Constantine, in the east of the country, Gérard wrote in his autobiographical book *The Life and Adventures of Jules Gérard, the Lion-Killer*, about his concern for the 'enormous losses which the lions inflict upon the tribes'.

He pondered if there could be some means of 'protection more efficacious than my own individual arm' that could save livestock from lions and be profitable at the same time both to Algeria and France. Perhaps the best way to seek out and kill the lion was to initiate an organised large-scale hunting project like that used in France to eradicate the wolf, Gérard mused.

Gérard then informed the French government that he and the local people were so keen to eliminate lions in the area

that they would foot an annual bill for 20,000 francs to pay
for such an operation, since this was roughly what lions were
costing herders in terms of devastating yearly livestock losses.

Although Gérard's grand proposal was not taken up by the
French authorities, plenty of lions were killed in smaller hunts.
Writing for the journal *International Zoo News*, Nobuyuki
Yamaguchi, associate professor of animal ecology at Qatar
University, noted in his article about the extinction of lions
from North Africa, that 202 lion kills were recorded in French
Algeria between 1873 and 1883 – and for these a bounty
'equivalent to £400 sterling was paid'.

Bryden also notes that as well as hunting, lions in Algeria
were also affected negatively by the more careful tending
of livestock that made killing them more difficult and an
even scarcer chance of hunting deer, which were becoming
increasingly rare.

In Morocco, the country's last few lions that had retreated
to the safety of the mountains found their sanctuary threatened
by armed bandits that moved into the hills. This, alongside
an increase in the availability of firearms and more generous
bounties for lion kills, led to more intense persecution and
saw the Barbary lion population dwindle into single numbers,
until eventually no more roamed the North African wild.

In the south of the continent, the arrival of the Dutch and the
British from the 1600s was catastrophic for the big cats of the
Cape. Just like the British in India and the French in North
Africa, the colonists brought guns, rifles and ammunition
galore, and set their sights on decimating as much wildlife in
a new land as they could.

Like the North African Barbary lion, the Cape lion had
two types of hunter to fear – farmers and big-game hunters.
For the Boer farmers of Dutch descent, Cape lions were
vermin that killed their cattle and lost them money. Whenever
lions were seen, they were shot down in pre-emptive strikes

to protect their livestock from future raids by the big cats. Lion after lion, potshot after potshot, wherever they were encountered, farmers destroyed them.

Writing in the 1700s about the Boers and lions, Anders Sparrman, a Swedish naturalist who travelled around the Cape between 1772 and 1776, said: 'The lion that has the boldness to seize on their cattle, is as odious to them as he is dangerous and noxious. They consequently seek out these animals, and hunt them with the greatest ardour and glee, with a view to exterminating them.'

Decades later, in the 1820s, Edward Turner Bennett, an English zoologist and author, described farmers' hatred towards lions in a similar way, noting that they were excellent shots who seemed to delight in killing lions. 'The Cape lion is seldom taken alive,' he said.

And in his 1959 book, *Spoor of Blood*, documenting the history of South African wildlife, author Alan Cattrick comments that hunting was an everyday activity for farmers, and that lions especially – which took their cattle – were vermin that must be killed at all costs.

There are also records of individual Boer hunters who killed large numbers of lions in the Cape. One such hunter called Petrus Jacobs is described by Guggisberg as a 'great lion hunter'. Cattrick adds that Jacobs 'probably eclipsed every [hunting] record that existed', having 'shot 110 lions and some 750 elephants'. Jacobs was also renowned for surviving a lion attack, aged 73, in which he had been badly 'chewed up'. Another Boer hunter called Kotje Dafel is also referenced by Guggisberg as bringing down 'more than a hundred lions'.

When British trophy hunters came into contact with the Cape lion, in particular the male, they were enamoured. It was love at first sight. Full of feline machismo, the lion of the Cape was a true he-lion, all muscle, meat and menace. He was a lion that made the hearts of hunters skip a beat. For them, he was a magnet, his head the ultimate trophy, the must-have head on the wall.

Much of the attraction was the lion's large size and dark-as-night mane. In *Travels in the Interior of Southern Africa*, published in 1822, English explorer and botanist William John Burchell, who arrived at the Cape in 1810, describes his first awe-inspiring encounter with a male Cape lion.

While exploring the north of the Cape in 1811, Burchell and his team set up camp north of the Orange River. As they inspected nearby reed beds, Burchell recalls how they suddenly became aware of a 'peculiar tone of bark … [that] … proved to be, lions'.

Burchell's dogs were sent in the direction of the roars, flushing out an enormous black-maned lion, and a lioness. The female sped back to the reeds, but the male headed towards Burchell and 'seemed to be preparing to spring upon [them]'. Before anyone could take aim, Burchell's dogs lunged forwards, barking ferociously and advancing to the side of the huge beast.

The lion responded calmly to the canine chaos with a 'majestic and steady attitude', swiping his paw, killing two dogs instantly. In response, every gun was instantly reloaded and fired at the lion, which returned to the reeds dying but 'with a stately and measured step'. Reflecting on the incident, Burchell wrote that the lion was considered to be 'of the largest size … as large as an ox' and with a copious mane that gave him a truly formidable appearance. This was a lion of a variety known locally as a black lion 'on account of the blacker colour of the mane, and which is said to be always larger and more dangerous than the other which they call the pale lion'.

Another visitor to the Cape, a Scotsman called Roualeyn Gordon-Cumming, was also besotted with the might of the male Cape lion. Described by Cattrick as 'the greatest hunter of his day', Gordon-Cumming arrived at the Cape in 1843 with the intention of 'penetrat[ing] into the interior further than the foot of any civilised man had yet trodden'. He hunted in South Africa for five years until 1849, later writing about his hunting exploits in a bestselling autobiographical book called *Five Years of a Hunter's Life in the Far Interior of South Africa*, published in 1874.

In the book, Gordon-Cumming details many lion hunts, but one in particular stands out because of the way he describes a Cape lion he has bagged. Waiting by a waterhole one day in September 1846, Gordon-Cumming reports how a lion with 'a most appalling roar' and 'a black mane which nearly swept the ground' suddenly came into view. Quick off the mark, the Scotsman took a shot, causing the lion to let out a deep growl before roaring mournfully as it retreated into nearby thorny bushes.

Following the lion into the scrub, Gordon-Cumming, with immense satisfaction, found 'the most magnificent old black-maned lion stretched out before [him]'. Prior to taking the lion's body back to camp for skinning, Gordon-Cumming lit a fire and gazed at his Cape lion booty, noting:

> No description could give the correct idea of the surpassing beauty of this most majestic animal … I … gazed with delight upon his lovely mane, his massive arms, his sharp yellow nails, his hard and terrible head, his immense and powerful teeth, his perfect symmetry throughout; and I felt that I had won the noblest prize that this wide world could yield to a sportsman.

Also hunting lions in the Cape around the same time as Gordon-Cummings was General John Bisset, a British Army officer who came to South Africa as a child in 1820. In his memoir, *Sport and War*, published in 1875, Bisset writes about his hunts for Cape lions as if they are military campaigns. In one chase for an enormous lioness, he records how it was 'necessary to call a council of war, to appoint a captain of the hunt' and pursue a single lion with a team of men some twenty-five strong.

After killing the lioness, Bisset describes her as 'a glorious trophy of our sport', so heavy that it took twenty strong men to lift her into the wagon. Once skinned, Bisset proudly had her pelt shipped to 'His Royal Highness the Duke of Edinburgh'. Bisset also describes the honour of taking His Royal Highness Prince Alfred hunting in the Cape in 1860. However, he notes disappointedly that although their 'first sport was ... to have been a Lion', the big cats remained elusive and they were forced to give up on the 'monarch of the plains ... for less noble game'.

It was Bisset who, in 1865, shot down the world's last Cape lion in Natal. Bisset was typical of the lion hunters in the Cape, equipped with high-velocity rifles and having the time for hunting, keen to send home magnificent skins. The combination of the persistent and prolonged fire from Boer farmers mixed with the fresh zeal of trophy hunters was deadly. From the early 1800s to the 1860s, in the space of around sixty or so years, the lions that had reached further south than any other lion on the continent were gone. Under so many lines of fire, the Cape lion just buckled and went over and out for ever.

Beyond the Cape, lion populations elsewhere in Africa were also plummeting as European colonists and settlers established themselves around the continent. In East Africa from the 1890s, when cattle farming became a large and serious business, lions were numerous and were regarded as a frightening nuisance.

Aggrieved by lions taking their livestock, farmers, just like the Boers in South Africa, considered them vermin and hunted them ruthlessly. Most were shot, but Jackson points out in *Lion* that 'traps and poison were also employed' as part of the 'concerted efforts ... to exterminate the lion'.

Some farmers were so plagued by lions that they employed professional hunters to rid them of the beasts. In 1911 Lord Delamare, a British peer and early settler in Kenya, recruited Paul J. Rainey, an American hunter, to destroy the lions that were attacking sheep at his farm in Soysambu in the Rift Valley. Rainey did such an excellent job that the *New York Times* reported that lions were 'being slaughtered like American rabbits' after hearing that he had killed twenty-seven lions in just thirty-five days.

It wasn't just farmers who employed hunters to kill lions. They were also commissioned by Kenya's Game Department to control lion numbers in the Masai Mara where, according to the environmentalist Bartle Bull in his book *Safari: A Chronicle of Adventure*, published in 1988, an 'overpopulation of lion ... developed as the Masai herds increased', leading to the hunting down of 'eighty-eight lion ... in three months'.

It was common for professional lion hunters to use dogs to help them bag their quarry. Rainey was a big advocate of using dogs, informing the *New York Times* that the best way to hunt was with dogs since they 'take the charge out of the king of the beasts and make the pastime more pleasant when the final death scene is enacted'.

Rainey kept a mixed pack of lion-hunting dogs at his farm at Naivasha, near the Kenyan capital Nairobi. He used his own bear hounds, shipped over from America, at the start of the hunt to course lions, followed by a pack of thirty or so mongrel fighting dogs to finish the lion off.

In South Africa, Gordon-Cummings also favoured hunting lions with dogs, using a mixed pack too – greyhounds for the start and wild mongrels for finishing. He used dogs so frequently that over the course of five years he lost seventy to lions.

Airedales and bull terriers were also used to hunt lions and were favoured by Roger Hurt, a lieutenant colonel with the British Army who hunted in the Rift Valley around Lake Naivasha. Hurt proclaimed that this combination of dogs was more effective than using Rhodesian ridgebacks. Also known as African lion hounds and African lion dogs, Rhodesian ridgebacks were bred in Rhodesia specifically to hunt lions. However, Hurt maintained that Airedales were faster than ridgebacks and bull terriers more aggressive, making them better dogs to hunt with.

Although dogs bred and trained to hunt lions made the chase more likely to close with a kill, there were some hunters who opposed the use of dogs. In *Safari*, Bull explains that hunting lions with dogs was considered 'unsporting' by some hunters because shooting lions distracted by dogs was far from skilful or fair.

With the construction of the railways, which started in the later 1890s, men were also employed to destroy lions that roamed the tracks and station taking track layers, engineers and station staff. One such lion killer was a young man, appropriately called John Hunter.

He arrived in Africa from Dumfriesshire in southern Scotland in 1905. His first job was as an armed guard working for the Uganda Railway, which linked Uganda with neighbouring Kenya. In *Safari*, Bull describes how Hunter would lean out of a train window looking for lions and leopards. When he spotted one, he'd then take aim with his old .275 Mauser. If he killed one, Hunter would then 'pull the alarm brake' and leap out of the train to skin the fallen beast. In time, as Hunter became more experienced, he trained the conductor to 'toot the whistle twice if he saw the lions and three times for a leopard'. Able to sell lion skins for £1 each, Hunter soon took up the activity full-time.

The Uganda Railway also had trouble with lions when it was being constructed, especially at Tsavo in southern Kenya. In 1898, while a bridge was being built over the Tsavo River, 135* labourers lost their lives to lions. For nine months, from March to December, men sleeping in the workers' camp were terrorised by a pair of male lions. While they slept, the two lions would visit, stalking the tents until an opening was found. Then they would crawl in, grab a sleeping man, drag him out and take him away to be eaten, feet first.

Precautions to deter the lions had been put in place. Camp fires were lit to scare the lions and thorny fences (bomas) built around the camp to stop the lions from getting in. Unfortunately for the labourers, the man-eaters were not afraid of the flames and either leaped over or crawled their way through the boma. In charge of the camp was Lieutenant Colonel John Henry Patterson, an Irish-born British soldier and engineer, who was also an experienced hunter.

---

* This is the number supplied by Lieutenant Colonel John Henry Patterson who was in charge of that section of railway. The Uganda Railway put the figure of dead men at twenty-five.

Yet despite his hunting experience and prowess, the lions evaded his shots, escaped from traps and never touched the poisoned carcasses carefully left out for them. Eventually, after twenty weeks of using the camp as a larder, Patterson finally got lucky and killed the lions. The first one was shot on 9 December and the second one on the 29th. Patterson had the pair skinned, using them as rugs until he sold them to Chicago's Field Museum in the United States some twenty-five years later. The man-eaters are still at the museum, displayed as reconstructed stuffed specimens.

The lions became notorious, their man-eating reported in newspapers of the time and even mentioned in the British Parliament. Patterson later wrote a book about his experiences called *The Man-Eaters of Tsavo*, which in turn inspired the making of three films: *Bwana Devil*, released in 1952; *Killers of Kilimanjaro*, released five years later in 1959; and *The Ghost and the Darkness*, which aired in 1996.

Tsavo wasn't the only section of the Uganda Railway to attract man-eating lions while under construction. At the Voi to Taveta line section in south-east Kenya, an engineer called O'Hara was, according to Guggisberg, dragged from his tent and swiftly eaten in 1899.

The following year, in June 1900, a male man-eater started visiting Kima station on the Nairobi to Mombasa line. It took and killed a labourer and then went after an engine driver who managed to evade the lion's jaw by hiding in a water tank. Keen to rid the station of the lion, Superintendent Ryall, a member of the Railway Police, attempted to put an end to the lion's visits by shooting him at night from a carriage in the sidings.

Ensconced in the carriage with two friends, the trio soon fell asleep. However, one of Ryall's chums, called Hübner, awoke and discovered 'to his horror that the lion was in the carriage. He had pushed open the unlocked sliding door.' Guggisberg quotes Hübner's description of what happened next after the lion pounced on Ryall's bed: 'This was the end of my friend! The right paw hit the left side of the head; the teeth sank deep into the chest ... Everything was deadly quiet ... The lion then dragged the body off the bed and deposited it on the floor.'

Eventually the lion made its exit by smashing through a window, taking Ryall's body with him. A sum of £100 was offered by the railways for the lion. He was caught alive, 'exhibited for a few days and a photograph taken before a rifle bullet put an end to his infamous career'.

Lions often took to roaming around the station houses built in lion country. When they appeared, station masters would send telegrams warning staff and passengers of their presence. In 1909, a newspaper in New Zealand called the *Manawatu Daily Times* picked up on these and published a small story featuring a series of telegrams sent by an Indian stationmaster at Simba, a little wayside station on the Uganda railway. They ran as follows:

Simba, 17.8.05. 1hr. 45 mins. The Traffic Manager: Lion is on the platform. Please instruct guard and driver to proceed carefully and without signal in the yard. Guard to advise passengers not to get out here.

Simba, 17.8.05. 7hr. 45 mins. The Traffic Manager: One African injured at 6 o'clock by lion, and hence sent to Makindu Hospital by trolly. Traffic Manager please send cartridges by 4 down train certain.

Simba, 17.8.05. 16hrs. The Traffic Manager: Pointsman is surrounded by two lions while returning from distant signal and hence Pointsman went on top of telegraph post near water tanks. Train to stop there and take him on train and then proceed. Traffic Manager to arrange steps.

Another cause for the decline of lion numbers was the growing popularity of the safari. As well as dodging the bullets of angry farmers and professional hunters, lions were now in the sights of sportsmen, happy to travel across the world, join a safari and bag themselves a lion at their leisure.

The rise of the safari is attributed to former American president Theodore Roosevelt, who visited East Africa as part of a shooting expedition to collect specimens for the Smithsonian Institution in Washington and the American Museum of Natural History in New York. Roosevelt left New York by steamer in March 1909 and arrived in Kenya a month later. The expedition finished in Khartoum in Sudan a year later.

An entourage of hundreds, including scientists, naturalists, trackers, guides and porters, accompanied Roosevelt. During the expedition, Roosevelt shot 512 animals, including 17 lions. Overall, around 11,400 animal and plant specimens were taken back to the United States, taking staff at the Smithsonian Institution 8 years to catalogue.

Even though Roosevelt was no longer president, his adventures in Africa attracted the world's attention. Newspapers ran stories covering his departure from New York and his arrival at ports of call in Naples and Cairo before his disembarkation at Mombasa in Kenya to begin his expedition.

While on safari, Roosevelt's caravan was shadowed by the Associated Press, who were committed to giving their readers the speediest coverage of what was to be known as the greatest safari ever. Their man in the field was former military and policeman Captain W. Robert Foran, who wired his stories via the Uganda Telegraph line, ensuring his news about Roosevelt always broke first. All over the globe, people read articles and features about Roosevelt's endeavours and were kept up to date with his latest activities by news film footage. Roosevelt effectively became the world's first safari star.

As the most orchestrated and publicised hunting expedition, Roosevelt's safari set a precedent. With such clamour, it didn't take long for those that could afford it to travel to Africa, sign up for a safari and start shooting. Hunting for big game was no longer the exclusive pursuit of British diplomats and colonials. Soon they were joined by entrepreneurs, millionaires and playboys, as well as aristocrats and fashionable types, with plenty of time, money and weapons to hunt down and shoot as much African wildlife – especially lions – as they liked.

To make the most of their time in East Africa and bag as much wildlife as possible, the new hunters required the services of experienced hunters – men who were remarkable shots and knew where to go to find trophies in land still home to lions and other great beasts. What they wanted were the services of a 'white hunter', who as Jackson notes in *Lion*, could be 'hired to guide wealthy clients on a hunting expedition and make sure they acquired enviable trophies'.

Fortunately for these wealthy clients, there were experienced hunters, already based in East Africa, ready and willing to rebrand themselves as white hunters. Some were professional hunters, making a living killing lions for farmers, landowners, railway companies and game departments. Whoever they worked for, they were expert shots, capable of taking out many lions in a short space of time.

One man who fitted the bill was John Hunter, mentioned earlier, who worked as a guard for the Uganda Railway, earning extra income from the skins of lions he shot from the trains. When he realised that he could make a good living as a professional hunter, he left the railways to offer his services elsewhere, including to wealthy safari clients.

Writing about his days as a white hunter in his book *Hunter: The Startling and True Adventures of One of the Last and Greatest of Africa's Professional White Hunters*, published in 1952, Hunter comments that some of his clients were aristocrats, including a French count and his wife who, following the fashion for big-game hunting experiences, 'wanted a few African trophies to decorate their chateau'.

Other white hunters of the day included the Australian Leslie Tarlton, with a reputation as a fine rifle shot, and his British colleague R. J. Cunninghame, described by Roosevelt in his book *African Game Trails* as 'exactly the type of hunter and safari manager one would wish for'. It was Cunninghame and Tarlton who organised Roosevelt's 1909 safari and were his chief guides.

Also part of the white hunter clique, with reputations as crack shots and with long waiting lists for their services, were Denys Finch Hatton, the second son of the Earl of Winchelsea, famous for his bravery when confronted by lions and who took the Prince of Wales out shooting in 1928; and the gentleman Philip Percival, whose clients included the author Ernest Hemingway and the immensely wealthy Baron Rothschild.

From Sweden, there was also the dashing aristocrat Bror von Blixen-Finecke, who ran a firm of safari guides and was described as a magnificent hunting companion. He was married to *Out of Africa* author Karen Blixen* from 1914 to 1926.

As the 1900s entered their first decade, another threat to lions emerged in the form of the motor car. 'Even more than dogs and high-powered rifles,' writes Bull in *Safari*, 'the car changed the balance'. Used as a substitute for chasing lions on horseback, camouflaged cars and jeeps could move quickly and travel great distances. Lions didn't stand a chance.

Even though they got stuck in mud and broke down, cars meant hunters could carry more equipment and ammunition, as well as more dead lion bodies quickly back to base. So effective was hunting with vehicles that in *Simba*, Guggisberg reports that up to sixty lions were killed in one lion hunt in the Serengeti 'thanks to use of cars'.

This scaling up of unsporting slaughter caused concern among some hunters. Percival, one of the white hunters, wrote in private notes that with the arrival of cars, lion kills 'doubled in half the time' but with the odds so unfairly stacked against lions, there really was no hunting skill involved in chasing lions by car.

Fellow white hunter Finch Hatton wrote a letter to *The Times* protesting about the 'orgy of the slaughter' at the Serengeti, caused by motor cars. Published on 21 January 1928, the letter noted the increasing numbers of visitors to

---

* Pen name Isak Dinesen.

East Africa 'who [were] anxious to fill their bag as quickly as possible', and that most of them cared little how they achieved this and the effect this might have on lion populations.

There's a photograph taken in Kenya in 1911 that shows what looks like the aftermath of a massacre. A group of hunters, trackers and porters, around thirty of them, stand with their rifles and spears in front of nine lions. All the lions are dead, laid out on the ground, limp and lifeless. It's a grim and heartbreaking scene made all the sorrier for the title of the image: *Morning's Work – Killed in 15 Minutes.*

All nine lions, one possibly a male, had likely been systematically shot, one after the other, execution-style – a pride blasted away in minutes. Glorious for the hunters, devastating for the lions. It is an image, albeit a grisly one, that holds up a mirror to the times.

It was a time when lions in Africa seemed so abundant that shooting down at least eight female lions, some of them quite possibly pregnant, in fewer than fifteen minutes was perfectly legal, an act to be proud of, to be photographed and championed. But for lions there was worse to come. Even darker times were just around the corner.

CHAPTER FOUR

People Hate Lions – Part II

People hate lions. The people who live with them anyway.

Craig Packer, in 'The Truth about Lions'
by Abigail Tucker, *Smithsonian* magazine

When we think of Africa, we see a wild place of vast grassy plains, uninterrupted vistas where the world's greatest migrations play out. A place where unfettered wildebeests in their tens of thousands kick up huge red dust storms as eagles criss-cross the sky above.

We see jittery herds of delicate gazelles and stately antelopes grazing on succulent grasses, and giraffes too, stretching their necks, reaching for the juiciest leaves from the top of tall acacia trees. And watching them all, through half-shut eyes, are prides of slumbering lions, giving in to the heat of a sunburned afternoon.

This golden panorama conjured up in our collective imaginations is a mirage. It's a vision of Africa past. Of a wilder, more abundant Africa, a storybook Africa. An imaginary snapshot of the East African landscape that captivated European colonists and settlers, immortalised in films like *Out of Africa*.

This was a habitat lions once thrived in. A landscape with few humans, where hundreds of thousands of herbivores, impalas, greater kudus and other antelopes were only threatened by the carnivores they lived alongside. This was a balanced ecosystem, only ever jeopardised by periods of extreme drought.

Today, landscapes like this barely exist. The wild, never-ending savannahs, the huge herds of antelopes and the lions that hunted them are vanishing, disappearing before our eyes. Outside of national parks and protected areas, pristine wild land is becoming increasingly hard to find. It's a point drawn attention to by Luke Hunter, president and chief conservation officer of the wild cat conservation organisation Panthera, in an article about the decline of African lions, published in *The Guardian* in December 2012, where he remarked: 'The really big, productive, well-watered savannah that are well protected outside of the national parks are a complete fallacy.'

Hunter made his comment after the results of a study assessing the amount of viable lion habitat left in Africa were published in a 2012 paper called 'The Size of Savannah Africa: A Lion's (*Panthera leo*) View' (also referred to in Chapter Two). In summary, the research, led by Jason Riggio, found that the 'the extent of savannah Africa' had 'shrunk considerably in the last 50 years', engulfed by agriculture, broken up into grazing land, fields and farms. He also noted that in most African countries where lions still range, the human population has quadrupled since the 1960s.

Land that was once lion country had been converted to agricultural use because of unprecedented human population growth. According to the United Nations, the population in Africa increased from almost 229 million in 1950 to 1.256 billion in 2017. As the number of people has rocketed, so too has the demand for crops and livestock, and the land to grow them on. Although urbanisation in sub-Saharan Africa is increasing, most of the population still lives in rural areas. Approximately 60 to 70 per cent of the population's livelihood depends on agriculture and livestock.

In his 2010 paper 'Managing the Conflicts Between People and Lion: Review and Insights from the Literature and Field Experience', written for the United Nations, Philippe Chardonnet, currently director of the International Foundation for the Conservation of Wildlife, noted that between '1970 and 2000, the area devoted to agriculture in sub-Saharan Africa increased by 25 percent'.

Illustrating the growth in the numbers of people occupying former savannah land, Riggio, in 'The Size of Savannah Africa', compared the number of people living on it in 1960 to the increased number some forty years later: 'In 1960, 11.9 million km^2 of these savannahs had fewer than 25 people per km^2. The comparable area shrank to 9.7 million km^2 by 2000.'

As populations have grown, they have taken over the wild landscapes of yesterday's Africa, grabbing larger and larger parcels of the lions' kingdom. It is now occupied territory. Savannah has been slashed and burned to create room for livestock, crops, farms, homes and towns. Some wild habitat losses have been dramatic and are most vividly illustrated in West Africa, where in Burkina Faso, for example, 80 per cent of the country's wild land no longer exists and in neighbouring Côte d'Ivoire, 79 per cent of the natural land has been taken.

In the last century, lions in Africa have lost 75 per cent of their former range. Savannah in Africa has shrunk so intensely that it is now more threatened than the world's tropical rainforests. The lion's des res is all but gone. It's this loss of land, the diminishing of their habitat, that presents one of the biggest threats to their survival that lions have ever known. But it's not just lions and other wildlife that have been forced from the land to make room for food and cash crops; traditional pastoralists have been affected too.

Unable to access land to graze their cattle, herders have moved their families and livestock to protected areas, often inside or on the borders of national parks, to continue grazing their cattle and establish new settlements. At the W–Arly–Pendjari Complex, a protected area of West African national parks that straddle Benin, Burkina Faso and Niger, 15.5 per cent of protected savannah is grazed by livestock.

Most of the continent's remaining lions survive in protected areas, so when herders establish themselves there, they inevitably bring their animals closer to lions. As cattle and goats, sheep and sometimes camels graze, the land degrades,

affecting wild herbivores that are now forced to compete
with livestock for grass. As the vegetation that herbivores
need to survive vanishes, they vanish too. As a consequence,
the prey base of lions is dangerously depleted. With less prey
to hunt, hungry lions turn to livestock instead. It's close and
available, the perfect fast food for ravenous prides.

In the east, in Kenya, lion habitat has also been lost because
of growing urbanisation. In the capital, Nairobi, homes,
offices, roads and other infrastructure have been constructed
at speed to accommodate its rapidly growing population. As
building land has run out in the city, conurbations have
spread to areas around the Nairobi National Park, the
country's first national park. Six kilometres south of the city,
just an electric fence on one side of the park separates it from
the metropolis.

When the park was established in 1946, it was surrounded
by wild land that allowed migratory animals like zebras,
wildebeests and gazelles, to move from Nairobi to the Masai
Mara in the south-west and Amboseli National Park towards
Mount Kilimanjaro on the Tanzanian border.

In the old Africa, migration was a simple affair. Obeying
their instincts, herbivores followed the rains in search of
fresher, greener pastures. Travelling the same migration
highways used by generations of animals, year after year, was
instinctive, safe and straightforward. Today, migrating from
Nairobi National Park is not so easy.

Migration highways are blocked, congested with people,
livestock, smallholdings and villages. Inevitably, as lions have
followed migrating herbivores, they've come across cattle and
goats kept by pastoralists and farmers. As lions have killed and
eaten them, they have come into conflict with the herders
they belong to.

The impact of domestic animals being killed by lions is
devastating for pastoralists and small farmers. For rural
Africans living in some of the world's poorest and least
developed countries, losing livestock – often their only form
of income – brings already impoverished families to their

knees. Lions are a large and dangerous predator to have to live with.

For some herders, dependent on just a couple of goats or cows, the consequences can be life-changing. The financial impact varies from country to country, but according to research cited in *Beyond Cecil: Africa's Lions in Crisis*, a report summarising the plight of Africa's lions published by Panthera and WildAid (an organisation dedicated to ending the illegal trade in wildlife), a single lion costs Kenyan ranchers at the Tsavo East National Park £223/US$290 per year in livestock losses. In other places, such as the Waza National Park in Cameroon, where 'lions kill cattle disproportionately compared to other less valuable livestock', lions were responsible for around 3.1 per cent of all losses 'but were estimated to represent more than 22% of financial losses – about US$370 [£285] per owner'.

Chardonnet also cites some financial examples. In Gokwe in central Zimbabwe, he notes that livestock losses between 1993 and 1996 averaged around '12 percent of annual income per household', and that of all the wild animals that predate domestic animals 'lions were responsible for the greatest economic losses'. For cattle herders grazing livestock around the Pendjari National Park in north-western Benin, the 'average annual loss to large carnivores such as lions was estimated at US$365 [£282]'.

Considering that the average income per person in Africa is estimated at around £1,591/US$2,004 per year or £4.36/US$5.60 per day, and considerably less for rural Africans, these financial losses are hard to bear, especially if there is no adequate compensation available to help mitigate them. Often farmers have no savings and are unable to replace the animals killed by lions that may have been their only source of income.

The resentment felt by pastoralists towards lions that kill their livestock, threatening their livelihoods and the financial survival of their families, can be overwhelming. They want to stop lions killing again and punish the culprits.

So they take justice into their own hands. Lions are quickly sentenced to death and are speared, poisoned or shot as soon as possible.

Ilkeek-Lemedung'I is a Maasai village of homes made from mud, stone and corrugated iron, established on the edges of the Nairobi National Park around 40 kilometres from the city. For most people living there, their goats and cattle are everything. So, when early one morning in June 2012 villagers awoke to lions attacking their animals, it felt like the sky was falling down.

Rather than take matters into their own hands and attack the lions, they called the Kenya Wildlife Service (KWS) to come and remove the big cats. The KWS duly arrived but without their vet, who was needed to tranquillise the lions so they could be taken back to the park. Uncertain when the vet would arrive and unable to watch their livestock being attacked by lions any longer, the Maasai took action to prevent more of their stock being killed.

As a result, two lionesses, two juvenile males and two cubs were killed. All the KWS could do was move the six dead lions, stained with their own blood, into their truck and take them away. Later the KWS reported that the number of lions present in the Nairobi National Park had now fallen from forty-three to thirty-seven. The incident made the press. AllAfrica, an African online news portal, reported the story, noting that the number of Maasai livestock recently taken by lions 'brings to 35 the number of sheep and goats and two cows killed by the lions from the Nairobi National Park in the last three days'.

Another incident that was picked up by the media occurred in the Manyara Region of northern Tanzania. There, on the night of 1 January 2015, six hungry lions made their way from the Tarangire National Park to a Maasai settlement around 96 kilometres away. Coming across a poorly built stable, they found and killed a number of donkeys. As news of the attack spread, six Maasai herders, armed with a gun and bows and arrows, set off after the lions. By dawn, all six lions were dead and four of the herders were injured. The Australian news outlet ABC covered the story and noted that such incidents were becoming increasingly common as herders and lions came into closer contact with each other, with locals 'launching vigilante operations against the felines'.

As well as using spears and guns to kill lions, poison is also used. Pesticides that are toxic to lions are easy to get hold of and are relatively cheap. Knowing that lions scavenge, angry herders sprinkle the dead bodies of livestock with the odourless granules of powerful poisons like carbofuran. Within a few hours, lions that feast on laced carcasses are dead, as are other scavengers, including jackals, leopards, hyenas and vultures.

In Kenya, on 7 December 2015, the poisoning of eight lions at the Masai Mara National Reserve hit headlines around the world. Well known to people in the UK, as well as locals, tourists and scientists, the poisoned lions were part of the Marsh Pride, made famous by TV presenter and zoologist Jonathan Scott and wildlife cameraman Simon King in their popular and long-running BBC *Big Cat Diary* programmes that followed the dramas of life in the pride like a soap opera.

Some of the pride had eaten a dead cow laced with lethal insecticide put out by local pastoralists in retaliation for killing their livestock. As a result, two lionesses and a young male died. Other members of the pride that had also eaten the toxic meat only survived because of veterinary intervention. Eleven vultures were not so lucky and also died after eating the poisonous carcass.

What was especially shocking about the poisoning of the Marsh lions was that they were persecuted inside the protected reserve area where it's illegal to keep livestock. However, under the cover of darkness, tens of thousands of cattle that graze beyond the reserve during the day are herded across the fenceless line that separates the reserve with its long grasses from overgrazed grassless pastures.

The zebras, elands and wildebeests that once grazed in the area are long gone, unable to survive on the degraded land. With the vanishing of these herbivores, the Marsh lions had taken to hunting and killing domestic cattle instead. Writing about the incident on his blog on the same day as the poisonings, Scott said the relevant authorities had 'turned a blind eye' to illegal grazing at the reserve and that in the last few years assaults against lions were escalating as numbers of cattle and their herders in the reserve increased.

A few days later on 9 December, Adam M. Roberts, chief executive of the Born Free Foundation, commented about the lions' deaths on the organisation's website, noting that the poisonings were 'devastating', predicting that if something wasn't done soon 'the terrible prospect of a Masai Mara without lions will become very real'.

The following year on 9 August 2016, three other well-known lions were poisoned in retaliation for killing a donkey in the Kunene Region in north-west Namibia. The lions were brothers and had featured in a film called *Vanishing Kings: Desert Lions of Namib*, which had aired on the National Geographic Channel earlier in the year.

There had originally been five brothers in the group, known as the Five Musketeers. However, one of them had been shot and killed a few months earlier, leaving just one lion surviving after the poisonings. He was swiftly translocated to an area less affected by human–lion conflict. Because of rising tensions with local people after a spate of livestock killings, the big cats had been scheduled to be relocated in a matter of days, just before the fatal poisonings.

All the lions were being studied and monitored by a conservation group called the Desert Lion Conservation Project. Aware that the lions hadn't shown recent signs of movement via their radio collars at the time of the poisonings, the project shared their concern on their Facebook page, posting the following: 'The three adult male lions ("the Musketeers") that moved 12 km into the mountains north of Tomakas suddenly stopped moving and their satellite collars went off-line yesterday morning.' The remains of the three lions' bodies and their collars were later found. They had been burned in an attempt to hide the evidence of their demise.

As the growing human population has encroached further into lion country, it's not just livestock that lions kill. It's people too. In many African countries, notably Ethiopia, Mozambique and Tanzania, man-eating lions present a serious problem. Lions enter villages and appear on front porches, fall through straw-roofed homes, pull sleeping grandparents out of bed and grab adults and children working in fields. For many people the sound of lions roaring close by is terrifying, and the likelihood of being carried off in the night or attacked in the bush feels frighteningly possible.

In Tanzania, where almost half of the world's remaining lions live, they have lost large swathes of their territory. With the human population tripling in just under 40 years, now edging towards 60 million, more than 37 per cent of Tanzania's woodlands are gone, most of them converted into farmland. Here, there are more human casualties as a result of encounters with lions than anywhere else on the continent.

According to Chardonnet, more than 120 Tanzanians are attacked by lions every year. He also cites the results of studies from 2005 and 2006 which note that 'lions killed 563 people between 1990 and 2006, and injured at least 308'. In southern Tanzania, south-west of the country's former capital Dar es Salaam, another study found that between 2002 and 2004 'at least 35 people had been killed in 20 months by one or more lions'.

Mozambique has the second-highest number of losses of people to lions. Since 1974, studies and reports state that at least seventy-three lion attacks on humans have been recorded at the Niassa National Reserve in the north of the country. At least thirty-four people have died and thirty-seven suffered injuries. Between 2004 and 2010, eleven people were killed and seventeen injured. Half of the attacks have involved lions entering villages and dragging people from their huts. Thirty-four per cent of incidents occur in the fields and in the bush. Between 2000 and 2001, seventy people lost their lives to lions in the north of the country in Cabo Delgado, as did a further forty-six between 2002 and 2003 in the Muidimbe district in the south-east.

The situation is different in West and Central Africa, where lion attacks are perceived as rare events. One of the last attacks in the region was recorded in 2007 and involved three villagers who were seriously injured by lions at the W National Park in the west of Niger. Because lion attacks are unusual, Chardonnet comments that people in the west are less fearful of lions because attacks are infrequent, and that 'nomadic and migratory pastoralists are generally not afraid of lions because they know how to frighten lions away with sticks and shouts'.

Although the slaughter of livestock is the biggest driver of conflict between people and lions in Africa, the fear of being murdered or injured by a lion is another reason why people kill big cats. In an article published in *Smithsonian* magazine in 2010, author and journalist Abigail Tucker writes about conflict between people and lions in rural Tanzania. She notes that in the past when a lion attacked someone, villagers would shoot only the nuisance lion, but now, using poison, whole prides are taken out in one go.

When it comes to human life and protecting loved ones, people don't want to take chances. Killing just one lion doesn't stop others coming into villages and wreaking havoc. If you live in a hut made of mud with just a flimsy thorn fence around it for protection, the threat of a lion breaking in is terrifying, and if, as Tucker writes in her feature, your cows are your livelihood and your little boy is a shepherd who sleeps outside with his goats, 'wouldn't you want to eliminate every last lion on earth?'

The psychological impact of lion damage to livestock and people can be greater than the economic hardships endured.

Deeply negative feelings towards lions, even in places where very few lions remain, can become entrenched. In many rural areas, lions are considered to be deadly vermin that must be eliminated to protect both people and livestock, even when the threat from lions is very low.

A letter published in the *New York Times* in August 2015 written by Goodwill Nzouaug, a student from Zimbabwe studying in North Carolina in the south-east of the United States, illustrates how the grim reality of living with lions can lead to their being viewed as 'objects of terror'.

In his letter, Nzouaug reveals that when he was a child a wild lion roamed around his village. After the lion killed people's livestock, local children were told not to walk to school alone and to stop playing outside. His mother was so frightened of the lion, notes Nzouaug, that she would only venture into the bush for firewood when accompanied by his father and older brothers, 'armed with machetes, axes and spears'.

After his mother's uncle was attacked by the lion, fear of the big cat intensified. Everyday life was so affected that 'no one socialized by fires at night; no one dared stroll over to a neighbour's homestead'. The lion, said Nzouaug, 'sucked the life out of the village'.

When the 'fearsome beast' was eventually killed everyone was delighted. No one cared who ended the lion's life. Whether it was killed legally or illegally, by a 'white trophy hunter' or a local poacher, was of no interest. What mattered to the villagers was their liberation from fear of the lion, a freedom celebrated by singing and dancing. With the lion gone, normal life was able to start again.

Nzouaug ends his letter with the news that recently a teenage boy in a village not far from his own was 'mauled by a lion and died' while he slept in a field. A potent reminder of the dangers that ordinary people living with lions face on a daily basis.

Likewise, a feature published in February 2017 on the CNN website detailing what life is like for people living with lions at the Masai Steppe in north-east Tanzania, revealed the negative feelings people can hold towards the big cats. Wildlife Warrior Lucas Lengoje was reported as saying that 'people are becoming more and more angry towards wildlife' and that they 'care more about their livestock than wildlife', largely because every cow lost to lions is a loss that many families just can't sustain.

Research has also been published regarding the negative attitudes towards lions held by people that live alongside them. For example, a 2000 study by Margaret Dricuru from the Uganda Wildlife Authority published the responses to a questionnaire sent to people living in and around the Queen Elizabeth National Park in south-west Uganda. One hundred and fifty-six questionnaires were completed and returned. The results showed that local people had little regard for lions. Thirty-seven per cent of respondents stated that the best way to deal with lions that strayed into their village was to kill them. A fence was suggested by 35 per cent, while 28 per cent of respondents thought the best approach was to learn how to avoid lions. Overall, attitudes towards lions were formed by interactions with the big cats, including loss of livestock and people, which generated fear about being near lions.

As part of his 2003 PhD thesis about conflict between people and animals, Hans Bauer, now working with WildCRU, the conservation research unit at Oxford University, questioned 236 people living in 10 villages on the edges of the Waza National Park in northern Cameroon about their perceptions of lions.

Of this group, 50 per cent expressed hostile attitudes towards lions. As a result of such negative feelings, lions are more severely 'punished' than other predators that kill livestock. For example, a study published in 2008 showed that in the southern Kalahari Desert, 'farmers responded lethally in 85 percent of cases of lion predation but only 55 percent of cases involving other large carnivores'.

The exact number of lions killed by people in retaliation for the loss of livestock and human life is unknown, although it is estimated that one hundred lions die as a result of conflict in Kenya every year. Killing lions is illegal in most African countries, so the incidents are not reported and are underestimated.

While the retaliatory killing of lions has been catastrophic, it isn't the only threat that lions encounter as they share diminishing amounts of space with humans. There are also dire repercussions for lions when people kill their prey in the form of illegal bushmeat hunting.

Hunting for bushmeat is not a new pursuit. It's been part of many African cultures for millennia, but in the past it was small-scale and sustainable. Today it's more like an industry. As well as feeding rural communities, bushmeat is sold to people in towns and cities, not just in Africa but in the United States and Europe, with an estimated 5 tonnes shipped to Paris every week.

For many rural Africans, buying bushmeat is an economic necessity and often the only affordable source of protein. For example, in parts of rural Mozambique a chicken can cost around £3/US$4. A similar-sized portion of bushmeat costs just £1.23/$1.60. Bushmeat is a lifeline that keeps many people from going hungry. In southern Africa, it's reported that around 6 per cent of the population relies on bushmeat, and in the Central African Republic that percentage rises to 55.

Although the need for cheap meat creates the demand for bushmeat in rural communities, in urban areas bushmeat is considered a luxury, with people said to be willing to pay 50 per cent more than they do for chicken and fish.

Meeting the demand for bushmeat has caused a massive decline in the prey that lions need in order to survive. Animals like wildebeests, zebras, buffaloes, impalas, kudus, giraffes and waterbucks have all been affected. Talking to *Scientific American* magazine in 2015 about the scale of bushmeat hunting, Luke Hunter summarised the situation as 'a commercial endeavor, not just a subsistence one', noting that bushmeat 'ends up on refrigerated trucks on highways and sold in cities'.

Originally bushmeat hunting took place in forests, but as they emptied, hunters moved out to the savannahs. Just as the forests lost their animals, so too have the great savannahs, even those in protected areas and national parks, notably in Angola, Cameroon, the Central African Republic, Côte d'Ivoire, Gabon, Malawi, Mozambique, Tanzania and Zimbabwe, where poachers have caused lion populations to decline and in some cases become extinct.

In the last thirty years or so, populations of wild herbivores have plummeted by 85 per cent in West Africa and by 52 per cent in the east. In the west, bushmeat hunting has long been a concern for conservationists, but now the impact is felt in the eastern and southern regions of the continent, too. In the Serengeti, for example, tens of thousands of wildebeests are now killed every year by poachers.

Guns, snares and gin traps are all commonly used by poachers to bag their meat. But it's snares that do the most damage. They wreak havoc. Easy and cheap to make, snares catch anything and everything, from endangered species to pregnant females, and, although intended for other animals, the deadly devices catch lions, too. Attracted by the smell of animals rotting on snares, many lions find themselves ensnared by the same deadly wire necklaces. Most lions die, slowly and horribly, usually by bleeding or choking to death; a fate that befalls 52 per cent of lions at the Niassa National Reserve in Mozambique.

In January 2017, the results of a Panthera study assessing the effectiveness of protected areas (PAs) were published in the journal *Biological Conservation*. Led by Peter Lindsey, research affiliate and director at Panthera, the study team analysed scientific surveys associated with 186 PAs across 24 African countries and assessed their ability to protect and provide habitat for lions.

Unfortunately, the study found that most African national parks and reserves were unable to cover their running costs and require stronger management capacity. As a consequence of this, PAs were not able to adequately protect lions and other wildlife. Regarding the funding of PAs, the report noted that levels of support are so low in some places that without intervention 'an increasing number of PAs will fail to fulfil the objectives for which they were gazetted'.

Such PAs are known as 'paper parks' – there on paper, but not fit for purpose, unable to provide robust law enforcement to counter poaching or resist political and social pressure to turn over their land. PAs, especially in Ethiopia, the Central African Republic, Angola and Mozambique, are losing land and lions because of chronic underfunding. The study concluded that across the continent, funding needs to 'increase 3–6 fold' to allow affected PAs to do their jobs properly.

Although not on the scale of the trophy hunting that drove the Barbary and Cape lions to extinction, wild lions in Africa are still killed in the name of sport. In twelve countries, including Benin, Burkina Faso, Cameroon, the Central African Republic, Mozambique, Namibia, South Africa, Tanzania, Zambia and Zimbabwe, it's perfectly legal – for the price of an expensive permit – to shoot lions and have their skins, heads and paws shipped across the globe for display in cabinets and trophy rooms at home.

Trophy hunters travel great distances from all over the world, but predominantly the United States, to kill lions in Africa. Between 2005 and 2014, the Humane Society of the United States estimated that American hunters shipped approximately 5,600 lion trophies home to the States. But it is the lion killed in Zimbabwe by an American hunter, Walter Palmer from Minnesota – who reportedly paid around £45,000/US$50,000 for the privilege – that everybody knows about.

On 1 July 2015, outside the perimeter of the vast Hwange National Park in west Zimbabwe, Palmer took aim at Cecil using his state-of-the-art compound crossbow. Palmer's arrow hit Cecil, but his shot wasn't fatal. Although injured, Cecil managed to evade Palmer and his guide. Determined to kill the lion, Palmer and company tracked Cecil for twelve hours. Eventually Cecil was found and Palmer finished off the thirteen-year-old lion with a bullet.

Before taking photographs, Cecil's collar was removed from his corpse.* Palmer then positioned himself next to the lion, his bow and arrow placed on Cecil's side to show, beyond doubt, that he had killed and now owned the once mighty lion spread out lifelessly before him. Then, according to *Oxford Today*, Cecil was skinned and beheaded.

What Palmer didn't know was that the lion he had just killed was well known in the area. Big, handsome and with a jet-black mane, Cecil was easily recognised and admired by regular visitors to Hwange. He was also very familiar to scientists based at the park. Fitted by them with a radio collar and given the name Cecil, he and his pride had been studied and documented since 2008 by researchers working with the Hwange Lion Research Project, as part of a study into wild carnivores managed by WildCRU.

Not only had Cecil, a beloved lion, been killed, but he had probably been shot illegally. Some reports say that Palmer's guide had lured Cecil from the park, where he was protected, with a dead animal tied to their vehicle. Unable to resist the carcass, Cecil left the park, never to return again.

News of Cecil's death spread. Photographs of Palmer and the lion went viral on social media. Much of the world was outraged. A new debate about the rights and wrongs of trophy hunting was ignited. People wanted to know how, when Africa is losing its lions at an unprecedented rate, the killing of further lions by trophy hunters could possibly be legal or

---

* Palmer later stated that he was unaware Cecil had been lured outside of the park and had not seen the lion's collar.

justified. Surely lion hunting should be banned to protect lion numbers from diminishing further?

These are emotive questions that aren't easily answered. Opinions about the rights and wrongs of trophy hunting vary among conservationists but fall broadly into two camps. Those against trophy hunting maintain that the sport contributes to the decline of Africa's already imperilled lions, is a threat to the survival of the species and should be banned. However, other conservationists say trophy hunting can, when managed and regulated properly, actually help to protect the continent's lions from further decline.

In December 2016, WildCRU director David Macdonald published his *Report on Lion Conservation with Particular Respect to the Issue of Trophy Hunting*, which had been commissioned by the British government. Its task was to consider how the UK might best support lion conservation in Africa. In the report, Macdonald – who is professionally neither pro- nor anti-trophy hunting but personally finds the activity disagreeable – states that the biggest benefit trophy hunting provides to conservation is the protection of wild lion habitat.

Currently, trophy hunting areas cover 1.4 million square kilometres of land, an area 22 per cent larger than that covered by Africa's national parks. If trophy hunting is banned, Macdonald argues, some of the land it operates on will be lost to the lion estate, most likely to agriculturalists rather than photo tourism providers, which will present 'a serious threat to the survival of the lion population'. Across much of Africa, MacDonald says, the conversion of trophy hunting land to agriculture is likely, either because it is 'profitable for land-owners, or in the case of government-owned land, because there is strong political pressure to turn these areas over to livestock production'.

To further illustrate his point, Macdonald includes an example of what happened to land after trophy hunting was banned in Botswana in 2014. After the hunting blocks were put up for sale, Macdonald reveals, there was 'minimal uptake ... for photographic tourism', and that as of June 2016, even though ten of thirty-two land concessions had 'been offered to public tender four times since the hunting ban', only one plot was operating as a photo tourism provider, 'the other nine being without concessionaires or not operational'.

Considerable concern has been raised about the future of these hunting blocks because they are part of a wildlife corridor that links the Moremi Game Reserve, Nxai/ Makgadikgadi Pans National Park, Chobe National Park and Hwange National Park (in Zimbabwe) together. Conservationists are anxious that if the corridor is fragmented by agriculture, not only would 'possibly the geographically largest intact lion population in Africa' be threatened, but so too would other wildlife, including buffaloes, elephants and zebras, which use the route ways.

Other areas of lion habitat particularly at risk if trophy hunting is banned, says Macdonald, are those in West and Central Africa. In the west, for example, just 250 lions remain as the result of habitat encroachment by people and livestock. The last two populations of any size survive at the W-Arly-Pendjari Complex, and Macdonald's review of the literature suggests that they only hang on there because the complex is 'buffered by hunting zones', which stop people moving onto the land. When land is protected by trophy hunting operators, lions are protected too. Without trophy hunting, lions would lose vital habitat, which would 'worsen the species' already deteriorating status'.

Other conservationists agree with Macdonald. Panthera president Luke Hunter, who personally finds trophy hunting 'repellent' but believes the sport can help protect lions, told *The Guardian* in 2011 that if hunting is banned, it's very likely that African governments – to generate revenue – would turn habitats over to crop growing and cattle raising.

Similarly, Guy Balme, director of Panthera's African leopard conservation projects, told *The Guardian* in July 2015 that if trophy hunting land was sold because of a ban, it would instead be farmed and 'accelerate the loss of wildlife'. Although trophy hunting is far from ideal, it currently 'slows the rapid decline of populations', making the activity 'a necessary evil'.

In addition to the preservation of wildlife habitat, some conservationists maintain that when managed ethically, trophy hunting can generate vital funding for conservation initiatives. For example, some hunting revenues are used to buy or rent land that otherwise might be used for farming. Just a relatively small number of lions need to die at the hands of trophy hunters for many more to be protected, helping safeguard the species as a whole, say proponents.

Trophy hunters tend to be wealthy individuals, willing and able to pay extraordinary fees – up to £85,000/US$108,000 in some instances – to buy a permit that allows them to shoot a male lion. Although there are fewer visiting hunters than tourists, each trophy hunter spends more money than a regular tourist does, pumping over £156 million/US$200 million per year into the African economy.

According to a 2012 paper called 'The Significance of African Lions for the Financial Viability of Trophy Hunting and the Maintenance of Wild Land' led by Peter Lindsey, lion hunting makes up 17 per cent of the continent's hunting income. Of all Africa's wildlife, lions generate the highest income per hunt.

It's not just lions on private hunting reserves that benefit from trophy money, say the pro-hunt lobby. Lions in national parks benefit too. Lyndsey also notes that most national parks in Africa receive inadequate state funding to help them protect their wildlife and are better supported by hunting operators, mostly for contributing towards anti-poaching activities. For example, Mozambique's Niassa National Reserve receives up to £314,000/US$400,000 per year from trophy hunting, almost 20 per cent of the funds needed to help protect its 42,000-square-kilometre area.

Using income provided by trophy hunting operators, the reserve has been able to remove thousands of snares and traps set by poachers, as well as to reintroduce lions. Every year, the Savé Valley Conservancy in south-east Zimbabwe receives £428,000/US$546,000 from trophy hunting operators and is able to employ 186 permanent wildlife rangers to safeguard its wildlife from poachers.

Trophy hunting money is also contributing to a 6-year rewilding programme (2017 to 2023) at the Zinave National Park in south-east Mozambique, where 6,000 large mammals, including elephants, giraffes, buffaloes and zebras, will be translocated from the Sango Wildlife Conservancy in a remote area in the east of the country.

The animals are being donated by Winfried Pabst, a German businessman, who bought Sango in 1993. The conservancy, although open to non-hunting visitors, is primarily a trophy hunting venue, where animals like lions, elephants and Nile crocodiles are shot and killed by paying clients. Trophy hunting revenue makes up 60 per cent of Sango's income. Without it, the conservancy land would be sold off and the 6,000 animals earmarked for translocation at Zinave would not be going anywhere.

Hunting money also funds some conservation research projects. These projects help scientists learn more about lion behaviour in order to develop initiatives that can minimise conflict with farmers and protect them from poachers.

In March 2013, Alexander N. Songorwa, the director of wildlife for the Tanzanian Ministry of Natural Resources and Tourism, wrote an opinion piece called 'Saving Lions by Killing Them' in the *New York Times*. In it he stated that trophy hunters spend up to 25 per cent more than regular tourists do, and also visit more remote places where they also spend money.

Every year, Songorwa said, around 200 lions are killed in Tanzania, 'generating about US$1,960,000 in revenue', in addition to money spent on accommodation, food, goods and travelling costs. 'All told,' said Songorwa, 'trophy hunting generated roughly $75 million for Tanzania's economy from 2008 to 2011' – money that supported twenty-six game

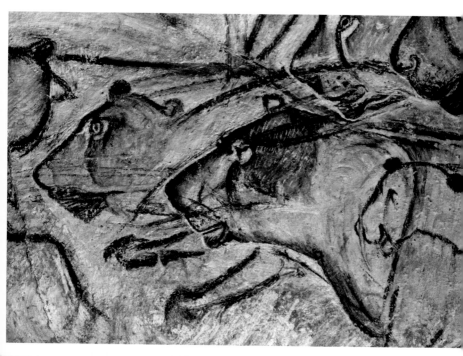

Above: Close up of the Panel of the Lions at the Chauvet Cave, France.

Left and below: Scenes from the Assyrian palace reliefs, showing the royal lion hunt of King Ashurbanipal (668–627 BC).

A Barbary lion in Morocco's Atlas Mountains, taken by Marcelin Flandrin in 1925 during a flight from Casablanca to Dakar. It is the last visual record of a wild lion in North Africa.

Taken in Kenya, this image shows the aftermath of a short lion hunt. It was part of a hunting expedition documented in the film *Paul J. Rainey's African Hunt*. The film ran in New York for 15 months, making it the 'most lucrative wildlife film of the period'.

Above: Lioness and cubs at the Linyanti Marshes in Botswana, as mentioned in the Preface.

Right: In northern Botswana, a lioness takes to the waters of the Linyanti Marshes.

Below: The serene lioness in Kenya's Masai Mara, as described in Chapter One.

Top: Resting male lion at the Kruger National Park in South Africa.

Above: Lions at the Port Lympne Reserve in Kent, thought to have Barbary lion heritage.

Right: A female Asiatic lion and her two cubs at the Gir Forest National Park, India.

This page: Some of the white lions at the Sanbona Wildlife Reserve in South Africa's Western Cape, as described in Chapter Two.

At the Selenkay Conservancy in southern Kenya, a Lion Guardian uses a radio antenna to locate a pride of lions.

Andrew Loveridge (left) and David Macdonald (right) from WildCRU take a break from their lion conservation work at the Hwange National Park in Zimbabwe.

Two of the seven sedated lions flown from South Africa to Rwanda for release at the Akagera National Park in June 2015.

Above left: One of the two male lions released at Akagera National Park in May 2017.

Above right: The first lioness to leave the acclimatisation boma and enter the wilds of Akagera National Park.

Next page: The lion known as Cecil at the Hwange National Park in Zimbabwe.

reserves as well as wildlife management areas owned and operated by local communities.

Trophy hunting supporters also point out that the activity provides employment opportunities – including working as wildlife guides and rangers, as well as in catering and hospitality – generating income for local people and economies. Not only does the extra money contribute to the financial welfare of local communities, but it also helps transform negative attitudes towards lions, say hunting advocates.

Where lions were once only seen as a lethal threat to both livelihoods and people, they are considered more favourably if they help put food on the table. Living lions become more valuable than dead ones, because when they're seen as a source of income, herders and villagers no longer want to kill them. Instead they're incentivised to help protect the lions they live with.

Defenders of trophy hunting point out that in areas where lion hunting has been banned, lion numbers don't always go back up. Sometimes they can actually fall and do so dramatically. For example, in Kenya, where a ban was established in 1977, the country's lion population was not saved. Before the ban was implemented there were around 20,000 lions. Today, there are fewer than 2,000. Hunting bans in Tanzania from 1973 to 1978 and in Zambia from 2000 to 2003 had a similar effect. Lions have instead been lost to shrinking habitats, retaliatory killings by farmers and as a consequence of bushmeat poaching.

Proponents also point out that trophy hunting is regulated to safeguard the long-term survival of lions and other hunted species. When run responsibly, strict quota systems ensure that the number of lions killed is sustainable. Guidelines suggest that a 'lion harvest' of between 5 and 10 per cent of a given population's males will not impact future population growth. In his report, Macdonald discusses quotas and notes that while these quotas may not be perfect, in countries where they are used responsibly, lion offtake is 'relatively conservative with no more that 2–4% of lions hunted annually'. For example, in Mozambique, Namibia and Zimbabwe, trophy hunting has not affected the population growth of those country's lions.

Responsible operators also ensure that hunters restrict their lion booty to mature males more than six years old. This is an easy rule for hunters to follow, hunting defenders say, since the nose tips of lions noticeably darken with age and are easy to spot. By targeting older males, the prides they leave behind are less disrupted. Populations are not impacted because mature males will have already spread their genes and sired the young of future generations.

When well managed and run responsibly, many conservationists believe that trophy hunting can play a positive role in lion protection, especially in terms of habitat preservation. However, it can also be a serious threat. When trophy hunting is poorly managed and run irresponsibly, lion populations suffer, particularly in localised areas where lions may also be under pressure from habitat loss, conflict with farmers and the effects of the bushmeat trade. When lions face these pressures simultaneously, it can be a deadly combination with hunting, as noted Panthera and WildCRU's *Beyond Cecil: Africa's Lions in Crisis* report, acting as 'an added threat to lion populations already under intense pressure from people'.

This scenario occurs at the Luangwa Valley in eastern Zambia. When lions remain in the South Luangwa National Park they are protected, but many wander out into the lion-hunting areas that border the park and are shot. Lions here are also at risk from angry herders who have lost livestock to lions, and from snares set by bushmeat poachers. A study at the park determined that all lion hunting in the area was unsustainable because of the other risks lions faced there, and that hunting should only continue if 'other forms of mortality could be significantly reduced, [because] any hunting at all [is] likely to contribute to population declines over a 25-year period'.

On this basis, advocates suggest that trophy hunting should only operate where all other risk factors are taken into account, and that where trophy hunting is permitted, it must be strictly regulated and ethically run so that revenues are shared fairly with local people and communities. On this basis, trophy hunting should continue until an alternative that doesn't affect wild habitats or local people's incomes can be found. Until that time, trophy hunting is a necessary evil, a bittersweet pill that can help protect Africa's lions from further losses.

South Luangwa National Park, Zambia

Out on a drive at the South Luangwa National Park, a zebra comes into view. It's a foal, a male. Gawky and unsure on his legs, he's by himself, separated from his mother and the rest of his herd. He's still shaky on the littlest of legs, all stripy, new and bony. He struggles to balance himself, faltering and slipping like he's on ice.

I'm not the only one to have stopped to look at the lost foal. A lioness has seen him too. From a tangled thicket behind the zebra, she comes charging, streamlined and intent on a kill. The startled foal looks around, his eyes huge and confused, his body paralysed by the fear coursing through his skinny veins.

The lioness skids into the foal. She lifts up a front paw and drops it down on him like a sledgehammer. Dust rises and falls as the foal sinks to the ground, pinned down under the lioness. Trapped beneath such weight, the foal can barely move. He lifts his neck and looks into the lioness's eyes. There are just whiskers between them.

The lioness stares back at him, her eyes disinterested, without expression. The zebra, his eyes still locked on the lioness as if trying to bond, folds back his black lips and brays quietly like an old donkey. It's a terrible, pathetic sound, a last

breathy whimper, maybe a call for his mother. All the while
the lioness looks casually around and then back at the foal,
taking her time, knowing she has her supper right where she
wants it.

It's excruciating – watching and waiting for the inevitable
execution. Fear has exhausted the foal. There's no way he can
move his frail body out from under the paws of a grown lion.
He lays down his head as if going to sleep. As he closes his
eyes, the lioness sinks her teeth into the foal's neck, snapping
fragile bones, tearing the tiny jugular open, draining out life
in a moment.

She clamps her jaw around the zebra's neck – scratchy bits
of baby mane catching in her teeth – and drags his carcass
back to the thicket she came from. There she can eat her meal
undisturbed, safe from other carnivores as well as the hunters
that look for lions outside the safety of the park.

In 2017, 'Lions, Trophy Hunting and Beyond: Knowledge
Gaps and Why They Matter' was published in the journal
Mammal Review. Led by WildCRU's David Macdonald, the
paper pointed out the difficulties involved in assessing
conclusively if trophy hunting is good or bad for lions. For
example, not enough is known about the causes of lion
mortality. The true extent of land used for trophy hunting is
unknown, as are the true finances involved. Likewise, no one
really knows what will happen if land used for trophy hunting
is sold off. 'There are extensive areas where the implications
of the removal of trophy hunting for lion conservation are
uncertain,' the research notes, 'because we do not know the
answer to questions like how much the industry's viability
depends on lions, or if lions could persist after an alternative
land use was substituted.'

So, while research into the true impact of trophy hunting
continues, the debate about whether the activity is really
good for wildlife and financially beneficial for African people

remains polarised. Jeffrey Flocken, regional director for the North American branch of the International Fund for Animal Welfare, which opposes trophy hunting, told *The Guardian* in 2016 that the fate of already vulnerable species should not be left to hunters on the as yet inconclusively proven basis that the benefits outweigh the drawbacks, when with 'trophy hunted species like lion and elephant populations crashing, we can no longer allow the killing of imperilled species for fun, based on blind faith that it's for their own good'.

Opponents of trophy hunting contest the premise that killing lions can help save them because of the revenue they make. They maintain that, on the whole, trophy hunting monies rarely reach local people and that any economic benefits received by African communities are so small they are negligible.

In support of their argument they cite findings published in a study called *Big Game Hunting in West Africa: What is its Contribution to Conservation?* published by the IUCN in 2011. It revealed that in countries where big-game hunting was legal, around 15 per cent of the land, some 1.1 million acres, was reserved for hunting.

However, despite the amount of land open to trophy hunting, the money made from the sport is only around 33 British pence/44 US cents per acre, a meagre 0.06 per cent of these countries' gross domestic product. In Tanzania, the study noted, the rate was even lower, making just 1.5 British pence/2 US cents per acre. The land used for hunting actually generates a lower income for local people than farming would.

Basically, rural communities receive next to nothing for turning over their land to hunting operators. For example, households in Zimbabwe with ten members might receive 77 British pence/US$1 to £2.30/US$3 every other year or so. In Tanzania, communities receive just £3.08/US$4 per square kilometre, even though the hunting operators are making £85/US$110 for the same size area.

In Kenya, tourism is reported as generating over £770 million/US$1 billion a year since hunting was made illegal in

1977. Hunting was calculated as operating at a loss of some £23 million/US$30 million annually. Since trophy hunting has stopped, the Kenyan economy has benefitted.

Likewise, in Botswana, conservationist and film-maker Dereck Joubert pointed out in a feature for *National Geographic* that after lion hunting was banned in the country ten years ago and Selinda Reserve in the north (previously a hunting reserve) was converted to photo tourism, its revenues increased dramatically – 'something like 1,300 per cent' – and described the increase in wildlife as 'spectacular'.

He also states that local people now receive training to develop and enhance their employability skills, and that previously hunting companies employed around 12 people, but now '180 people are employed full time, just in this one reserve'.

Employment opportunities are also over-egged by trophy hunter supporters, say opponents. According to the IUCN's study, trophy hunting generates only around 15,000 salaried jobs across Africa – a very low number considering more than 150 million people live in eight of the big-game hunting countries. Around 1,300 trophy hunting operators employ roughly 3,400 guides. Most operators have, on average, around 14.5 clients per year, with each guide working with 5.5 trophy hunters in a year.

When the economic rewards are 'a pittance', the IUCN report concludes, there is no incentive for local people to want to stop killing lions or to protect their habitat. When financial incentives are so poor, it's not going to change local people's mindsets and attitudes about protecting wildlife, let alone the lions they come into conflict with. There simply is no financial impetus to preserve habitat or prevent poaching. People earning perhaps 46 British pence/60 US cents a year are not going to stop poaching for an income that might buy them just one egg.

Another report published in 2013 called *The Economics of Trophy Hunting in Africa are Overrated and Overstated*, authored by Economists at Large, a team of independent economists specialising in environmental economics (commissioned by the Born Free Foundation and the International Fund for Animal Welfare, among others), drew similar conclusions, with the report's lead author Rod Campbell noting that 'the suggestion that trophy hunting plays a significant role in African development is misguided' and that revenues can 'only constitute a fraction of a percent of GDP', most of which is never passed on to local people.

In another Economists at Large report, *The $200 Million Dollar Question: How Much Does Trophy Hunting Really Contribute to African Communities?*, prepared for the African Lion Coalition in 2013, the widely quoted statement used to defend the role of trophy hunting in lion conservation: 'Hunters and hunting actually benefits Africa's lions – as well as its humans. Revenues from hunting generate $200 million annually in remote rural areas of Africa' is unpicked.

The quote comes from an article by Larry Rudolph and Joel Hosmer called 'Why Being Hunted is Good for Africa's Lions', published online in 2011 by the Daily Caller, a 24-hour news portal. The Economists at Large report states that Rudolph and Hosmer most likely sourced the US$200 million (£154 million) figure from a paper called 'Economic and Conservation Significance of the Trophy Hunting Industry in Sub-Saharan Africa', published in the journal *Biological Conservation* in 2007.

Economists at Large state that the estimate of hunting income quoted in the paper is not based on published, publicly available and reliable data. Their calculations also fail 'to employ any sort of systematic methodology'. More worryingly, half of their final revenue estimate is from a South African source that can't be traced. The only figures that can be easily found and show any scientific methodology come from Namibia and Tanzania, leading Economists at Large to note: 'The widely quoted $200m revenue estimate is based on very weak sources and methodology and should be used with caution.'

It's not just local communities that don't benefit from trophy hunting, say opponents – it's lions, too. Trophy hunting does not save lions; it just kills them and in the long term threatens whole lion populations because the number of trophy lions killed is unsustainable. Although there may be guidelines for sustainable lion hunting, the issue, say critics of trophy hunting, is that very few lion populations are regularly and accurately surveyed, so it's simply not possible to assess what level of take-off is safe.

Some countries, like Zambia and Zimbabwe, where trophy hunting is legal, have in the past witnessed declines so dramatic that hunting has been suspended temporarily to give beleaguered lion populations a chance to recover. Once moratoriums were in place, lion numbers began to bounce back, proving, say opponents, that the cause of the initial declines was over-hunting.

Too many lions are shot, leaving behind too few animals to breed to make up the shortfall because many quotas are simply set too high, say opponents. Some are set too high because of inaccurate population data, and some because of pressure from the hunting industry. The other challenge with quotas is that even when they are set sustainably, they are ignored by unscrupulous operators and hunters, either because providers who are known to turn a blind eye can attract more clients, or because hunters offer operators more money, off the record, to do so.

Julian Rademeyer, the author of *Killing for Profit*, a book about the poaching crisis in southern Africa, is quoted in an article about trophy hunting published in the *Los Angeles Times* shortly after the killing of Cecil the lion in Zimbabwe. Talking about the pressure that some trophy hunting operators are under to ensure their clients get the trophies they hanker after, Rademeyer says that unregulated and corrupt outfits result in 'a killing frenzy'.

In the same article, Craig Packer, director of the Lion Research Center at the University of Minnesota in the United States, says that his ongoing study of lion populations in Tanzania found over-hunting to be a crucial reason for their decline, largely because the government there failed to cut hunting quotas or introduce a law to protect immature male lions 'because of widespread corruption in the industry'.

In 2010, 'Effects of Trophy Hunting on Lion and Leopard Populations in Tanzania' was published in the journal *Conservation Biology*. A team of biologists from the University of Minnesota, led by Packer, looked at the causes of falling lion numbers in Tanzania. By analysing the number of lions brought back by legal trophy hunters, the team discovered that from 1996 to 2008, the number of lions killed by hunters fell by half. Next, they considered what could be causing the decline. They found it wasn't because there were fewer hunters or fewer hunting safaris. Far from it.

Evidence showed that since 1998 the number of hunting trips booked had increased by 60 per cent. Also, it was implausible that hunters were shooting fewer lions. After all, as Brian Child, a geographer from the University of Florida in the United States and not connected with the study commented: 'In general, if they're [trophy hunters] paying a lot of money, they're going to be hunting as hard as they can.'

The researchers also considered the other threats that lions face, including habitat loss, disease and retaliatory killings by herders. However, there weren't enough of these to account for such a drop-off in numbers. Rather, the decline was caused by legal trophy hunters who were responsible for 92 per cent of the fall in numbers. The hunters weren't breaking the law. The quotas set were just too high for lions to bounce back. Hunters were killing and bringing back fewer trophies because there were fewer lions left to kill.

Packer concluded that 'trophy hunting appears to have been the primary driver of a decline in lion abundance in the country's trophy-hunting areas' and that the only way to reverse the situation was for quotas to be lowered and the shooting of young males to be prohibited.

In their 2014 study 'Surveys of Lions *Panthera leo* in Protected Areas in Zimbabwe Yield Disturbing Results: What is Driving the Population Collapse?', researchers led by conservation biologist Rosemary Groom from the University of Johannesburg in South Africa, studied lion populations in hunting concession land around Gonarezhou National Park in south-east Zimbabwe and at the Tuli Safari Area, a controlled hunting area in the south-west of the country.

They found lion populations there to be much lower than anticipated. At Gonarezhou, where they had expected to find between 115 and 357 lions, they found only 33. At Tuli they found no evidence of any lions at all, even though data suggested the area could support between twenty-one and forty-two big cats.

The main cause of the decline in both areas, researchers ascertained, was 'unsuitably high trophy hunting quotas'. In Tuli, they learned that 'a total of 16 male lions [were] on quota between 2000 and 2009' and 'three on quota in ... 2006 and 2007'. These quotas, they noted, were unsustainable and were a 'high number for an area of only 416km^2'. These quotas had a significant impact and quite possibly wiped out the local population. Other reports studying the effect of over-hunting on lion populations in Hwange National Park (Zimbabwe), South Luangwa National Park (Zambia) and the Benoue Complex (Cameroon) have drawn similar conclusions.

Too many young males are shot, say trophy hunting critics, causing long-term threats to lion populations. When a pride loses its dominant males, everything changes. Prides without their male protectors are vulnerable. Outside males quickly move in on an unprotected pride and, keen to take over, will fight other males, often to the death, causing even more lions to be lost. Once new males have asserted their dominance, they will kill any cubs in the pride. When a female loses her cubs, she will return to oestrus, enabling the new male to sire cubs with her.

While this is natural behaviour, when too many males are being killed, the structure of lion prides and societies becomes disrupted with the repeated killing of cubs. When what Luke Hunter calls 'the essential mantle of protection that allows a female to raise a generation of cubs' is destabilised, lion populations unequivocally start to fail.

Taking only older males in a hunt minimises this impact, and many trophy hunting operations say their hunters only take aim at lions over the age of six. But, says Dereck Joubert, it is 'a myth that hunters only shoot old males past their prime that have been cast out from the pride'. Joubert also points out that older, cast-out lions 'almost always lose their manes from stress' and that battered, gnarled and maneless old lions are not the kind of trophy that hunters want to pay thousands of pounds to display in their homes.

Lions aren't just killed by trophy hunters and herders; poachers are killing them for their bones too. In Africa, lion parts have been used as decorations, as good luck charms and in traditional ceremonies in a number of countries, including Benin, Burkina Faso, Cameroon, Kenya, Nigeria, Senegal and Somalia, for thousands of years.

Lion parts are also ingredients for traditional medicine. Lungs have been used to treat children with whooping cough. Backache and joint pain have been eased with skins; veins have been used to restore sexual potency. Lion body parts and derivatives, mainly lion fat, are still used today in traditional medicine, but the trade is relatively small-scale, creating negligible impact on lion populations.

What's far more damaging is a new demand for lion parts, evidence for which began to emerge in 2009. The essential lion ingredients are not claws, skins or internal organs, but bones. Toe bones, back bones, tail or leg bones – it doesn't matter what bones they are as long as they are the bones of a big cat.

In many South Asian countries, including China, Vietnam, Laos and Thailand, drinking rice wine fortified with bones from wild tigers has long been popular because of the brew's perceived medicinal effect. Once consumed, it's believed the drinker will become strong and vital, like a tiger. Arthritis and rheumatism will be cured and erectile dysfunction becomes a complaint of the past.

In 1993, when the number of wild tigers reached critically low levels, China banned trade in their bones. This made it very difficult to make or buy tiger bone wine, but it didn't stop people wanting to drink it. To satisfy the demand, lion bones from Africa, in particular South Africa, began to be used as substitutes for tiger bones. From 2005 onwards, the labels on many Chinese medicines and wine bottles began to feature images of lions instead of tigers.

Selling lion bones in South Africa is legal. Brewers in Asia buy and import them from commercial lion breeders, whose main trade is supplying trophy hunters with lions that are easy to kill. Sometimes, bones come from trophy hunt kills, and sometimes lions, usually females, are killed specifically for their bones. A complete lion skeleton can sell for more than £3,000/US$3,847.

It's a trade that's on the rise. A study called *Bones of Contention: An Assessment of the South African Trade in African Lion* Panthera leo *Bones and Other Body Parts*, published in 2015 by TRAFFIC (a wildlife trade monitoring network) and WildCRU, reported that from 2008 to 2011, 1,160 lion skeletons – the equivalent of 10.8 tonnes of bones – were legally exported from South Africa, almost half of them in 2011 alone.

By contrast, before 2008 the United Nations Convention on International Trade in Endangered Species (CITES) had issued permits for just three lion skeletons to be exported to Denmark. However, seven years later CITES noted that there was now 'clear scope for the international trade in lion body parts for [traditional Chinese medicine and traditional African medicine] to grow uncontrollably, as it has done for other big cats'.

However, the report led by Vivienne Williams, a researcher in ethnoecology, ethnobotany and wildlife trade at the University of the Witwatersrand in Johannesburg, South Africa, found little evidence to suggest that the lion bone trade in South Africa was affecting wild lion populations, noting that 'the trophy hunting industry ... is the main source of carcasses once the trophy hunter has taken the skin and skull', rather than lions poached in the wild.

While the report notes that wild lions in South Africa are not threatened by the growing trade in their bones, it does suggest that the situation requires ongoing monitoring, especially since the market for bones has given value to female lions that previously had little or no worth to breeders of lions for canned or trophy hunting. The report also points out that while South Africa's wild lion population may not be affected yet, the continent's other wild lions, especially those in West and Central Africa and also in parts of Mozambique and Tanzania, may not be faring so well.

For example, in 2009 a man from Vietnam was arrested and then deported after securing lion bones and parts without official permits. Two years later in June 2011, two Thai men were found with fifty-nine illegal lion bones and were subsequently arrested. In April 2017, Earth, an online environmental news portal, reported that lion carcasses were being bought by wildlife traffickers at the Niassa National Reserve in Mozambique.

Officials from Niassa said the trade had been going on for around a year and a half, and that the carcasses were being bought for around £1,170/US$1,500 or traded for motorcycles. One reserve official was quoted as saying: 'There's very little local demand for lion parts … If it's happening at Niassa, it's supplying bones for the Asian lion bone market.'

There's also a specific demand for the bones of wild lions, as these are believed to create more potent wines and medicine than those made with the bones of captive-bred big cats. This is a concern – especially in West and Central Africa where underfunded national parks offer poor protection for lions against poachers.

Many conservationists fear that the illegal market for lion bones is growing and will soon affect Africa's last remaining wild lions. Sharing this view, Colman O'Criodain, a wildlife trade specialist for the World Wildlife Fund (WWF) was reported in *The Guardian* in December 2016 as saying that

WWF were 'very concerned' about the trade in captive-bred lion bones because it 'keeps demand for big-cat bones alive, and complicates enforcement efforts'.

While people with their weapons and snares are an obvious menace, lion populations have been placed in peril by another threat, one unseen and unheard as it approaches. In modern times, disease, in its many long-named guises, has been the killer of thousands of lions – with feline immunodeficiency virus, canine distemper virus and bovine tuberculosis being the most well-known and potentially devastating diseases.

Similar to the human immunodeficiency virus (HIV), feline immunodeficiency virus (FIV) compromises the immune systems of big cats. At first, lions with FIV don't show obvious signs of sickness and can appear perfectly healthy for years. But beneath the skin, the disease weakens lions' immunity to disease and infection. For example, lions with FIV find it harder to survive parasitic infections, including tapeworm and hookworm.

Lions pass on FIV by biting and during mating. Females can pass the virus on to their cubs. In Botswana, South Africa and Tanzania almost 100 per cent of lion populations in certain areas are infected. Other populations, including those in West Africa and at Namibia's Etosha National Park, appear to be free of the virus, although some scientists suggest they may have a strain of the virus not yet identified.

However, unlike domestic cats, which are likely to die from the disease, wild African lions, according to a related IUCN factsheet, show no 'evidence that FIV infection results in increased mortality'. The same factsheet also states that although FIV has only been recognised since the 1980s, lions may well have had the disease quite 'possibly for thousands of years' and developed immunity to it. Studies of infected lions at Serengeti National Park in Tanzania and the

Kruger National Park in South Africa have all shown that the disease is 'not a health threat'.

As a precaution, though, the IUCN suggests that lions which are FIV-negative, including those at the Etosha National Park in Namibia and Hluhluwe-Imfolozi Park in South Africa, should not be translocated to populations where lions do have FIV, because 'these lions may not be resistant to FIV induced disease as they have not had a chance to adapt to the virus during evolution'.

In 1994, in early February, tourists flying over the Serengeti National Park in a hot-air balloon at sunrise spotted the first lion believed to be suffering from canine distemper virus (CDV), a condition that affects digestion, breathing and neurological systems. The lion, a male, was seen twitching and convulsing, until eventually it collapsed on the dry earth beneath it. By sunset the lion was dead.

Within a year, around a thousand lions – a third of the park's population – had died the same way. CDV outbreaks weren't confined to Tanzania; they spread across the border to Kenya and into the Masai Mara where lions soon displayed typical symptoms of circling behaviour, muscle twitching, head jerking and involuntary movements of the jaw, known as chewing gum fits, as well as fits and seizures.

The source of the CDV was traced to the domestic dogs, around 30,000 of them, that lived in the villages dotted around the park's edges. More than likely, bat-eared foxes, hyenas and jackals first caught CDV from these dogs and then passed it on to lions as they mingled with the big cats at kill sites inside the park.

As human settlements around national parks grow, CDV remains a threat. Some dogs are vaccinated against the virus, but many aren't. It's a very real concern for the small populations of lions dotted around Africa, especially in west and central areas, as well as for lions in the Gir National Park in India where a small population of just over 500 live in an area 10 times smaller than the Serengeti. If just a few of these lions contracted CDV, it would be very easy, in just a short time, for the population to be wiped out altogether.

Bovine tuberculosis (bTB) was probably introduced at the close of the eighteenth century by infected domestic cattle brought into South Africa by settlers from Europe. From South Africa, it spread up into the grasslands of the Serengeti and woodlands of northern Tanzania and southern Kenya. BTB interferes with lions' digestion and absorption of nutrients, making affected animals emaciated. Many lions also develop problems moving when painful lesions form in their bones and joints. Coughing and biting among lions causes the disease to spread.

At the Kruger National Park in north-east South Africa, a significant number of lions have bTB, probably caught by eating infected buffalo carcasses. Every year, the park loses around twenty-five lions to bTB. In the southern range of the park, an estimated 48 to 78 per cent of lions are infected. Those in the north are clear, for now.

In 1999, a research project was initiated to compare the life spans of thirty-two lions – sixteen with bTB and sixteen without – in the north and south of the park. By 2003, twelve of the sixteen infected lions from the south had died: five from bTB and seven either because they had been killed by other lions or shot when they left the park, 'possibly as a result of the apparent disruption of the pride structure when the other individuals died of advanced bTB'. In the north of the park, eight of the lions in the study remained alive.

The impact of the disease doesn't just affect lions in terms of fatalities. It also affects their social behaviour. When adult males, which have the highest prevalence of bTB, become infected, they become weak. They're no longer able to defend their pride from healthy competitors and find themselves ousted. As the new dominant male takes over, he kills any cubs before attempting to sire some of his own, disrupting pride structures and creating stress much like the effect that the over-hunting of males can also do.

There are vaccine options to curb bTB in buffaloes, but vaccinating buffaloes is difficult and costly, and not widely adopted. Both bTB and CDV are considered hazardous to the health of Africa's few remaining large lion populations.

As people have moved into lion country, building and farming on the land, lions have been forced into ever-smaller patches of fragmented, often fenced-in territory. In such small, cut-off areas, male lions find it difficult to roam, to go beyond their pride and find unrelated females to mate with. The consequences of prolonged inbreeding are severe, eventually resulting in population bottlenecks that threaten the long-term survival of lions.

Inbred cubs often die in the womb. Cubs can be malformed and sickly, unlikely to reach adulthood because weakened immune systems make it difficult for them to fight off infection and disease. Sometimes, they can display neurological impairment and unnecessary aggression. This has been seen in captive lions that have not been bred using stud books that would help prevent inbreeding.

For example, in February 2014 the BBC reported that five lions – a lioness and her four cubs – had been 'destroyed' at Longleat Safari Park in south-west England because of 'serious genetic defects caused by inbreeding'. When the lioness arrived at Longleat in 2011, aged eighteen months, she displayed unusual behaviour that was attributed to dietary deficiencies. However, it was soon concluded that the lioness was actually 'suffering from neurological problems' as a result of 'relatively high levels of inbreeding'. Not only was the lioness affected, but so too were her cubs, displaying neurological abnormalities and aggressive behaviour that was 'not considered normal'.

The irreversible neurological disorders faced by the lions were so severe that, according to the *Irish Times*, the park said 'it would have been irresponsible to move the animals to another collection', and that it 'reluctantly took the decision to euthanise them'.

One of the world's most studied group of lions are an inbred population living at the Ngorongoro Crater on the eastern edges of the Serengeti Plain in Tanzania. The crater, a 610-metre-deep caldera – one of the planet's largest – covers around 162 square kilometres and is often described as 'a microcosm of African savannah'. Because of the crater's cliff-like walls and the presence of powerful resident males, lions living here are effectively marooned from other big cats that roam the neighbouring Serengeti. However, crater lions don't need to leave the area to follow migrating herbivores because the microclimate ensures that grasses grow year-round, providing prey for them, almost on tap. As a result, the lion population of around a hundred lions at the crater is one of Africa's densest.

In 1979, lion expert Craig Packer and fellow scientist Anne Pusey started a decade-long project to find out if the lions at the crater were inbred as a result of their separation from other lions, and, therefore, subject to vulnerabilities. Writing an article for *National Geographic* about the study, Packer explained the reasons for suspecting the crater lions were inbred. In the first instance, he noted that records suggested that no new lions had entered the crater in the previous four years, and secondly, that a report had indicated that in 1962 the lions in the crater had been decimated by a plague of biting flies, which would have affected the 'genetic richness of the group'.

Research led Packer and Pusey to conclude that all the lions on the crater floor 'descended from 15 animals'. Eight of them were survivors from the biting fly plague. The other seven were new lions from the Serengeti that had been able to take advantage of the absence of healthy males caused by the plague, and to find themselves females with which to mate in the crater.

After the plague, life returned to normal and breeding resumed. Males in the crater then closed ranks, evicting any other new lions roaming in from the Serengeti, essentially locking down access and preventing new blood from entering the population. 'Thus,' said Packer, 'the population has been subject to close inbreeding since 1969, about five lion generations'.

To assess the genetic consequences of such inbreeding, leading geneticist Steve O'Brien from America's National Cancer Institute and colleagues from the National Zoo in Washington took and compared blood and semen from lions at the crater with samples taken from free-roaming lions at the Serengeti, where 'close inbreeding is almost non-existent'. Results from the samples revealed that lions at the crater were inbred and were also producing sperm that had levels of 'abnormality twice as high as the Serengeti males, another indication of inbreeding'.

Further test results brought more worrying news. They revealed a 'striking lack of genetic variability in the crater lions' immune defence systems', making the population more likely to be decimated during an epidemic. Results also showed that although the population in the crater had a good number of adult males, only a few of them 'fathered most of the offspring', meaning that the breeding population was actually smaller than expected.

This loss of genetic diversity could well be the cause of the low-level reproductive rates of the crater lions, suggesting that future reproductive rates may 'continue to decline ... unless new males are once again able to enter the crater'. Lions that are inbred are often infertile, nature's way of stopping unhealthy lions from reproducing. The quality of sperm from male lions can be severely compromised, resulting in no fertilisation, stillbirths and very small litters.

It's highly likely that other small, isolated populations in Africa are similarly inbred, vulnerable to bottlenecking and becoming locally extinct. Inbreeding particularly affects small, reintroduced populations. During the 1960s, three lions were brought to the Hluhluwe-Imfolozi Park in KwaZulu-Natal, South Africa. From the original 3 lions, a population high of 120 lions was reached. All were inbred, however, and this number fell to eighty-four and then crashed to just twenty between 2002 and 2004.

For lions to flourish they need genetic diversity, and for genetic diversity, large, free-roaming lion populations are required – something in short supply all over Africa. The effects of inbreeding threaten populations that have fewer than ten prides. A large and genetically diverse lion population is made up of fifty to a hundred prides. On this basis, Africa has only eight genetically diverse populations comprised of at least a hundred prides, found in Angola, Botswana, Chad, the Central African Republic, Tanzania and Zimbabwe. That's just six countries with lion populations not threatened by the consequences of inbreeding.

In his *National Geographic* article about the lions at the Ngorongoro Crater, Craig Packer comments that their history 'may represent the future for many other large vertebrates', since increased human presence around national parks has 'formed virtually impermeable boundaries'. This causes many species to become cut-off in small populations, making them vulnerable to disease as well as being affected by close inbreeding, forcing many small populations 'in a downward spiral'. Small lion populations not as well-known and well-studied as those in the crater will 'run their course undetected'.

Although Packer's words were written more than a decade ago, they're poignant because they remain pertinent, especially with regard to the small, mostly forgotten populations of lions in West and Central Africa, some as tiny as two in Nigeria and sixteen in Senegal. These populations are as isolated as the crater lions, and are unlikely to survive for much longer.

Lions in Africa are victims of a disappearing geography, of a continent being 'unwilded'. Without wild habitat, natural prey and adequate protection from pastoralists and poachers, the king of the beasts has been brought to its knees.

The threats lions face aren't going away. They are intensifying. In the next three decades, they will be exacerbated by climate change and continuing rapid growth of the human population in sub-Saharan Africa. The numbers are staggering. Figures from the United Nations predict that in Africa the population will grow from 1.256 billion in 2017 to 2,526 billion in 2050, leaping to 4,467 billion in 2100.

Inevitably, as the number of people increases, so too will the need for land to graze more livestock on. There are predictions for this growth. By 2050, the area of land converted for agriculture will rise to 51 million hectares, up 21 per cent from 2005, and the number of livestock will increase to 1.2 billion, compared to 688 million in 2005.

As the natural landscape transforms at supernatural speed and plains are replaced with grazing land, lions lose their natural prey and the wilderness they need to survive. They and Africa's other wildlife are set to be erased from the map, just like the savannahs. But not everyone is prepared to sit back and let this happen.

CHAPTER FIVE

People Love Lions – Part I

My soul is among lions.

Psalms 57:4, *King James Bible*

In his book *Monster of God*, which considers the relationship between people and carnivores, author and journalist David Quammen describes the west of India as a mainly dry and scorched region where the 'soil looks hard and weary' and 'lava-rock slabs lie naked in the sun'. But in one small patch of land, he says, just a few kilometres from the Arabian Sea, in the Kathiawar peninsula, there grows a rich and verdant forest that stands out from the rest of the 'generally sere and bedraggled continent'.

The forest Quammen refers to is the Gir Forest in the state of Gujarat. It's a forest that is home to treasure – gold, in the form of 500 or so Asiatic lions, the only wild lions living outside of Africa. In 1880, just a dozen were found here, the last survivors of the centuries-long hunting spree that wiped out the rest of the Middle East and North Asia's lions.

At the time, the future appeared dire for these lions. As they teetered on the edge of extinction, the world prepared to say farewell to Asiatic lions forever. The lions, though, had other ideas.

Almost half a century on, in 1920, a census recorded fifty Asiatic lions at large in the forest. In 1963, 285 lions were counted. Over the next 40 years, lion numbers went up in their hundreds, with 304 counted in 1995 and 441 in 2010.

By 2015, the most recent census showed the number of lions had reached 524, a 27 per cent increase in just 5 years.[*]

This sustained growth in the lion population was recognised by the IUCN, when in 2000 Asiatic lions were downgraded from critically endangered on its Red List of Threatened Species to endangered. Seventeen years on, however, and despite the growth in their numbers, lions in India remain listed as endangered because they still face a high risk of becoming extinct. The population is still frighteningly small and vulnerable, but conservation efforts in India continue to make a real difference in ensuring the lions' chances of long-term survival.

Historically, one of the reasons lions survived in the Kathiawar peninsula while other lions in India were wiped out by the 1870s was because of the area's remoteness and unforgiving dryness. Most of the region's terrain consists of hard and rocky soils, flat plains and a few small rivers, surrounded by hills and mountains covered in thorn and scrub. The landscape, with its combination of dry deciduous forest, inhospitable scrubland and steep slopes, was difficult to tame.

Elsewhere in India, in the mid–1880s, other regions were starting to industrialise. Huge tracts of forest and jungle were being cleared to make room for large-scale agriculture and to meet the growing demand for teakwood. Railways were linking states and moving goods around the country. As natural habitats were destroyed to make room for commerce, fields and plantations, animals described as 'venomous snakes and dangerous beasts' were exterminated in their thousands. Estimates suggest that 80,000 tigers,

[*] All census/population numbers for Asiatic lions living in the Gir Forest and other protected areas have been sourced from asiaticlion. org/population-gir-forests.htm. Retrieved 9 September 2015.

150,000 leopards and 200,000 wolves lost their lives between 1875 and 1925, with thousands more killed, their deaths not recorded.

At the Kathiawar peninsula, though, industrialisation and the large-scale growth of forestry and agriculture took longer to take hold. In his 2006 paper 'Junagadh State and its Lions: Conservation in Princely India, 1879–1947' published in the journal *Conservation and Society*, Divyabhanusinh, a conservationist and author, points out that one of the reasons the forests remained intact was because their teakwood was not deemed good quality, and so they avoided the 'pressures of timber extraction'. Cattle grazing, because of the poor nature of the region's soil, remained relatively small-scale and domestic.

Not only was the peninsula's terrain difficult to develop, its society was considered uninviting, too. Divyabhanusinh describes it as a 'buckaroo place, relatively unpopulated, politically unstable, a gladiatorial arena for opportunists' where, since medieval times, pirates, warriors and warlords had fought each other for various principalities, of which there were more than 200.

The Gir Forest itself was home to brigands, outlaws and rebels on the run from the region's various princes and chieftains. So treacherous was the area that historian and military man Captain Harold Wilberforce-Bell, in his 1916 book *The History of Kathiawad*, described it as a place where 'everything was chaos and confusion … robbery was rampant'. Plagued by malaria-carrying mosquitoes and many kilometres away from large centres of commerce, the peninsula was not top of anyone's list for development.

These factors enabled Kathiawar's wild animals to escape the systematic eradication inflicted on beasts considered dangerous elsewhere in India. The isolation of the peninsula with its untameable landscape provided safety for lions, first in the mountains and hills and then in Gir Forest, protecting them while the rest of their species perished elsewhere.

Although the lions of the Kathiawar peninsula may have avoided becoming the collateral of India's growing industrialisation, their numbers were still affected adversely by hunting, or *shikar*, as it was known. Mostly organised by local princes, lion shoots were popular, enjoyed especially by British and European colonials.

In 1880, Colonel James Watson, a British officer and keen hunter, undertook what is believed to be the first lion census at Gir. He estimated that 'not more than a dozen lions' lived in the forest. Considered a competent, trustworthy and knowledgeable man, Watson's findings, according to Quammen in *Monster of God*, 'alerted a few people among the British and princely houses that the Indian lion was not an infinite source of sport or nuisance ... but a precious rarity'.

Junagadh's ruling prince of the time, Nawab Mahabat Khanji II, who ruled from 1851 to 1888, was acutely aware of this. Conscious that the lions which had once roamed beyond Gir in the hills of Barda and Alech to the west and at Chotila and Dhrangadhra to the north were gone for ever, Mahabat Khanji II acknowledged that he was now the Asiatic lion's sole custodian and that the big cats' future rested in his hands.

Alarmed at the paucity of lions left in his forest and encouraged by the then British Governor of Bombay, Lord Sandhurst, Mahabat Khanji II asserted his princely right to control who hunted his lions. On 1 May 1879, he published a notice in the *British Gazette* stating that a royal interdict against the destruction of lions in the Gir Forest had been issued to prevent the lions becoming 'extinct in nature', and that this ban applied to everyone, including European sportsmen.

From then on, anyone wishing to hunt lions in the Gir Forest had to seek approval from the nawab to do so. Without official permission and a royal *diktat*, hunting Asiatic lions in the forest was forbidden. This ruling by the nawab was the first milestone in protecting *Panthera leo persica*, considered – according to Divyabhanusinh – to be the 'historical foundation of modern day conservation efforts' and the 'earliest initiative in the Indian Empire for protecting a species for its own right'.

Although Gir's lions now received royal protection, the precious few that were left were still being shot and killed. Mahabat Khanji II, despite his concern for the Asiatic lions living in his forest, was at heart a traditional royal. His ban existed largely to ensure that enough lions survived for him and his elite entourage to carry on hunting them, turning the big cats into increasingly rare trophies. The fact that so few lions now roamed the earth made them all the more desirable to hunt, more prestigious to kill, the ultimate limited edition big-cat booty to bag.

By now, the Gir Forest, with the outlaws it once harboured long gone, was effectively a royal hunting reserve, complete with luxury hunting lodges, offering a once-in-a-lifetime opportunity to shoot one of India's lions. Controlling who took part in such prestigious *shikar* enabled Mahabat Khanji II to demonstrate his superiority and authority, issuing select invitations and approving or denying requests to hunt lions from British officers and dignitaries, as well as other Indian nobles and royalty keen to take part.

This royal approach to protecting lions in order to carry on hunting them continued for decades – not just by Mahabat Khanji II, but by the nawabs that came after him, too. Rasul Khanji, the succeeding nawab of Junagadh from 1892 until 1911, shared Mahabat Khanji II's admiration and concern for the lions of the forest.

Although considered an expert marksman, skilled enough to 'take out a tossed coin with a shot', Rasul Khanji refused to shoot any lions himself. However, this didn't stop him delighting in royal *shikar*, issuing invitations to British and European dignitaries and other Indian royals to join his lion-hunting parties. Lavishly executed and with no expense spared, Rasul Khanji's shoots were events well-remembered by those privileged enough to take part.

Eight years into his rule in 1900, Rasul Khanji invited Lord Curzon, the British Viceroy of India from 1899 to 1905, to join one of his lion hunts in his forest at Gir. Lord Curzon duly travelled to Junagadh and met the nawab, but unexpectedly declined to join Rasul Khanji's lion-hunting party in the forest.

Curzon's refusal to join the hunt is thought to have been influenced by a local newspaper opinion piece he read in Bombay before he headed to the peninsula. The article pondered why, when there were so few Asiatic lions left, were members of the British aristocracy jaunting around the Gir Forest intent on killing the precious few big cats that remained?

Perhaps the piece struck a chord with the lord. Whether it was the newspaper feature that caused Curzon to stay away or something else, Rasul Khanji's response to Curzon's absence from the hunt is certain. In a letter sent to the Viceroy after he had left Junagadh, a dismayed Rasul Khanji wrote of his great disappointment in Curzon's abandoning of the lion shooting and pressed the lord to join him on another shooting excursion before his departure from India.

Unmoved by Rasul Khanji's wish for him to join another hunt, Curzon instead made his concern about the future of Gir's lions known. In 1902, he wrote to the Bombay Game Preservation Society, stating that those who believed India's lions were doing well were misguided:

> The facts seem to me to point entirely in the opposite direction. Up to the time of the mutiny, lions were shot in central India. They are now confined to an ever-narrowing patch of forest in Kathiawar ... The present generation owes it to its successors to restore the only species of a large mammal lost in historic times. Failure to do so would not be forgiven by the judgement of history.

Curzon was right to be worried. Although estimates concerning Asiatic lion numbers for the time vary – from around eleven adult lions in 1900, up to seventy in 1903 – the population was perilously low.

As well as regulating the shooting of lions in the forest of Gir, Rasul Khanji implemented other measures to help protect his big cats. During the early 1900s, severe drought badly affected Gir's lions. As their prey starved to death, unable to feed on anything but parched vegetation, lions were forced to leave the forest to find food elsewhere.

This led them to villages and settlements where they took cattle and goats, and sometimes people. As lions satisfied their hunger, the Maldhari – the pastoralists who lived in Gir – reacted violently. In retaliation and in self-defence, Maldhari began killing the nawab's lions. Although Rasul Khanji accepted that people had little choice but to slay lions to protect themselves and their loved ones, he did try to reduce the number of retaliatory killings.

As soon as the prince heard that lions were reported to be prowling close to villages, he would send men armed with burning torches and firecrackers, loud as thunder, to scare straying big cats back to their homes in the woods. To help minimise the anger felt towards lions, the nawab gave Maldhari and other pastoralists financial compensation for the loss of their livelihoods. Essentially, Rasul Khanji was mitigating conflict between humans and lions using methods still employed today.

While this helped safeguard lions in Junagadh, protecting lions that wandered into neighbouring states wasn't so straightforward. When lions left the Gir Forest in search of food, they often roamed into adjacent states, notably Baroda and Jetpur, where they killed both people and livestock.

Because the people affected didn't live in Junagadh, the nawab refused to offer compensation for lost livelihoods or give permission for his lions to be shot in self-defence. Instead, Rasul Khanji instructed them to adopt techniques, like those used by his men, to ward off stray lions and force them back into Junagadh. The residents of neighbouring states were not impressed by his terms and continued killing the lions they believed posed a threat to them. The nawab was so aggrieved by this, he wrote a letter to the Viceroy of India (then Gilbert Elliot-Murray-Kynynmound, the 4th Earl of Minto), complaining about the negative attitudes of his neighbours. In May 1909, he also issued a formal notification prohibiting 'the destruction of lions, lioness[es] or their cubs' in a bid to prevent the extinction of the species.

However, the notification had little effect. While the nawab refused to pay any compensation to those affected by his lions' behaviour or provide the big cats with food to keep them in the forest, the rulers of Baroda and Jetpur refused to guarantee the safety of any lions on their land.

While the prince had no luck convincing his neighbours to leave his lions alone, he was more successful when it came to protecting their home, the Gir Forest. As early as the 1870s, some twenty years before Rasul Khanji came to the throne of Junagadh, two-thirds of the Gir Forest was already gone – much of it had been illegally felled to make room for arable and pasture land. Some of the land was converted to industrial use, and some was used for building new homes for a growing population.

In response, Rasul Khanji and his advisers began to draw up plans to protect what forest was left. To this end, the nawab found support from some British colonials, who welcomed the initiative to save Gir's mighty forest. Before the 1850s, it had covered as much as 3,218 square kilometres.

One such colonial was Lord Charles Lamington, the Governor of Bombay from 1903 to 1907. After visiting the Gir in 1905, Lamington had been so impressed by efforts to preserve it that he 'deputed a forest officer from the Bombay cadre for preservation and development of the Gir Forest on scientific lines'. A year later, in 1906, Lamington praised Rasul Khanji's new plans to set up a lion sanctuary in the forest to protect the few lions that remained in India.

Two years later, in 1908, the sanctuary referred to by Lamington, the first of its kind in the British Empire, covering just over 321 square kilometres, was established. That same year, the forest was classified a reserve, and management of the forest was transferred from Junagadh's State Revenue Department to the State Forest Department. The forest area was increased too, with the State Forest Department managing around 1,528 square kilometres of forest, compared with just 885 square kilometres in 1906/7.

In 1911, Rasul Khanji died. His successor, Mahabat Khanji III, aged eleven, was too young to rule, so British administration was imposed on Junagadh until 1920, when the new prince came of age. During these nine years, protection for India's lions began to really mean protection. Up until Rasul Khanji's death, 'lions were being shot at the rate of four or five per year within Junagadh's portion of Gir and eight per year in the outlier regions'.

L. L. Fenton, a serving British field officer at the time, wrote in 1909 in the *Journal of the Bombay Natural History Society* that despite the protection offered by the nawab 'there cannot be the slightest doubt that [lions] are gradually, but surely, approaching extinction'.

To save the lions, the British administration issued an outright ban on hunting them. Now in charge of issuing hunting permits, all requests, including those coming from members of the British and European elite – as well as Indian royalty – were refused. Even when a request arrived via the Secretary of State for India in 1914 for an Asiatic lion to stuff and display at the British Museum in London, the administration stuck to its guns and informed the museum their answer was no, since permission would 'be liable to misconstruction of an undesirable kind even in the interest of science'.

For almost a decade under British administration, while lion hunting was forbidden, lion numbers began to increase from around twenty in 1913 to fifty in 1920. According to Quammen in *Monster of God*, 'the worst years' for India's lions 'had passed'.

In 1920, the heir apparent came of age and was invested as Nawab Mahabat Khanji III of the state of Junagadh, with full powers. This prince became known as a renowned – if eccentric – animal lover. He owned at least 300 dogs, each provided with its own room and servant. When his favourite dog, Roshanara, 'came of age', Mahabat Khanji III looked out for a husband for her.

When he found a suitable match in Bobby, a golden retriever owned by a nearby prince, the nawab pulled out all the stops to celebrate their impending nuptials. British officials and Indian royals were invited to the three-day-long ceremony. Roshanara travelled to her wedding in a silver palanquin and the groom was welcomed to Junagarh by a guard of honour, accompanied by 250 dogs dressed in fine brocade outfits.

Thankfully for Gir's lions, the nawab's love of animals extended to more than domestic species. The big cats living in his forests had a special place in his heart, too. Yet despite his affection for them, Mahabat Khanji III, like his royal forefathers before him, considered Gir's lions royal game. So, once he was in power, the hunting of lions for sport that the British administration had put a stop to started up again.

However, Mahabat Khanji III was less generous in issuing permits to hunt lions than Rasul Khanji had been. Records from the time show that in a twelve-year period during his reign, perhaps fourteen were shot and killed by dignitaries and royals with official permits. Unfortunately, records detailing the total number of lions killed this way during the whole of Mahabat Khanji III's reign are not available.

Although the nawab 'shot lions as was the custom of the time', he shot them 'sparingly'. Aware of their increasing scarcity, he shot just one lion during the period 1920 to 1933. When he visited the forest at Gir for royal *shikar*, huntsmen, in advance of his arrival, would lure out the largest male lions with the darkest manes for the prince to shoot from his *machan*, a special shooting platform.

However, the nawab would never take aim. Instead, Mahabat Khanji III would suddenly announce his plans to leave. It's said that the real reason the nawab attended these shoots was not to slay lions, but to keep an eye on his staff and keep them on their toes. Compared with other Indian royalty, Mahabat Khanji III's hunting habits were negligible. At this time, other nobles were killing hundreds of tigers, the Maharaja of Rewa, for example, singlehandedly killing almost 1,000 during the 1920s.

While the nawab was exercising restraint in his *shikar* and the granting of permits, he wasn't able to control the killing of lions that strayed from Junagadh. What happened to his lions in neighbouring states was beyond the limits of his royal powers and influence. Just like Rasul Khanji II before him, Mahabat Khanji III faced the same challenges with neighbouring royals regarding the protection of his lions.

The view of the princes living in the states next door was that, if lions roamed into their royal realms, then they gained the right to do what they liked with the beasts since they had, de facto, become theirs. According to them, Mahabat Khanji III lost his rights over his lions when they crossed borders, becoming fair game just like any other wandering animal.

It wasn't just that his few beloved lions were being shot that angered the nawab, though. It was the fact that many of them were being deliberately lured from the forest so they could be shot as soon as they left his land.

A year into his rule, in 1921, the nawab became involved in what Divyabhanusinh describes as 'a full blown lion crisis'. It was started by a letter received from the maharajas of Bikaner and Navanagar asking for a permit to hunt lions in part of the Gir Forest that did not fall within Mahabat Khanji III's royal boundaries.

The request riled Mahabat Khanji III. His response to the maharajas was a resounding no. Didn't they understand, the nawab pointed out, that in 1920 alone, twelve precious lions had already been killed outside of his boundaries, so, of course, even though they had asked for permission, he could not allow any more of his rare lions to be killed for sport?

The nawab also vented his anger in a series of letters sent to the then Governor of Bombay, George Lloyd. In them, he expressed his outrage at his neighbours' behaviour and attitudes and requested official support to stop the shooting of lions that ventured out of the Gir Forest. In his letters, the prince referred to the practice of his neighbours deliberately luring his lions onto their land by tying up buffaloes, describing such activity as 'poaching' and an activity that the British should be sorting out.

He then went on to note that without support from the British government, he might be forced to take matters into his own hands to protect his lions from being poached. And that, in doing so, affrays would no doubt be caused at state borders, the likes of which might destabilise and threaten the peace that the region presently enjoyed.

The governor was also asked to observe that the Gir Forest, most of which belonged to the nawab, was financially significant since it provided grass and firewood and also generated rainfall. Mahabat Khanji III also made clear his family's historic bond with the forest as the only refuge of India's last lions.

With no holds barred, Mahabat Khanji then went on to declare that if the British did not 'use their influence to see that [his] rights are respected and especially the scandal of tying up buffaloes within sight of [his] borders is stopped', then he would 'certainly adopt a policy of disafforestation'.

Despite threatening the peace of the region and destroying his own forest, which would have serious implications for the people and climate of the whole peninsula, the British administration remained unmoved by the nawab's passionate call for help in preventing the extinction of Asiatic lions. They offered no practical support, just a note that the 'complete disappearance' of the lions would be regrettable, but that this was as far as their sympathies went.

For the nawab, this was a terrible blow. Without involvement from the British, there was little he could do to protect his wandering lions. He had pulled out all the stops in his letters and had nothing left to negotiate or bargain with. The British had called his bluff and, of course, he couldn't go through with destroying his own forest, the Asiatic lions' only home.

Mahabat Khanji III, the lions' solitary custodian and guardian, was unable to prevent their being lured from the relative safety of the forest to face gunshot and death at the hands of his royal neighbours. Estimates of the number of lions whose lives ended this way vary, but counts suggest that between 1920 and 1943, close to ninety lions were shot and killed outside of Junagadh. Mahabat Khanji III did all he could, but without support, his efforts to protect his lions once they wandered from the forest were impotent.

As the 1930s dawned, over 6,000 kilometres away support for the long-term survival of Mahabat Khanji III's lions was growing. In the UK, naturalists and conservationists with links to India were acutely aware that the Asiatic lions living in the Gir Forest were on the brink of extinction.

In London, the Society for the Preservation of Fauna in the Empire (SPFE, now known as Flora and Fauna International), initiated a campaign to save the Gir Forest and its lions. The project was headed in London by Richard Onslow, the 5th Earl of Onslow, and backed in India by his brother, the then Viceroy of India.

They worked hand in hand with Mahabat Khanji III, establishing new rules and regulations to further protect the lions of Gir. All hunting in the forest, from lions to boars, was banned and a system of alerts set up so that lions straying from the forest into other states were lured back in, using buffalo and other meat as bait.

Negotiations were started with the nawab's princely neighbours to try and convince them not to shoot lions that wandered into their territories. Success came when the Maharaja of Bhavnagar was talked around, promising that neither he nor anyone in his princedom would shoot another lion. Some of the royals and landowners in the smaller states, including Jetpur and Bilkha, agreed to compromise, confirming that none of their citizens would be allowed to shoot or hunt Gir's lions. However, as royals, they reserved the right to continue indulging in *shikar* when the opportunity to bag one arose.

There was no luck with the maharajas of Baroda and Gondal, though. They would neither agree nor compromise on the issue. And so, lions that crossed over state borders and into their land remained fair game for anyone with a gun, spear or poison to hand.

To assess what else could be done to protect the world's last Asiatic lions, the SPFE commissioned Keith Caldwell, a respected game warden based in East Africa, to travel to India and study the lions at Gir. Caldwell was asked to determine the most pressing threats to the lions' long-term survival.

After spending time in the forest studying the lions, Caldwell concluded that the biggest cause for concern regarding the lions' future was the lack of natural prey for them to hunt and feed on. He calculated that for around 200 lions to have a long-term future, at least 8,000 to 10,000 prey animals, including antelopes, deer and boars, needed to live in the forest. Caldwell pointed out that the actual number of ungulates and other herbivores in the forest fell well below this.

Having observed that antelopes and other herbivores were extremely jittery and fearful at the approach of humans, Caldwell deduced that many of the animals were being illegally hunted for bushmeat, decimating the lions' source of natural prey. Caldwell maintained that if the numbers of forest ungulates could be boosted, then as a knock-on effect lions would be far less likely to look to livestock for their nourishment.

As fewer domestic animals would be taken from pastoralists living in the forest, the number of retaliatory lion fatalities would fall. In response to Caldwell's suggestions, Mahabat Khanji III agreed to ensure that all his hunting rules were adhered to, and that ungulates, just as much as his big cats, would be offered full protection from hunters and poachers.

In 1936, the first official and scientific survey recording the numbers of Asiatic lions was carried out by the Junagadh Forest Department. It estimated that 287 lions were living in the forest. This was a terrific increase in the population, when, according to an unofficial survey carried out by Sir Patrick Cadell (a respected member of the Royal Asiatic Society and an employee of the Indian Civil Service) in 1920, just fifty big cats were recorded.

Having one set of official numbers was a conservation milestone. Although surveys had been carried out in the past, they were informal and numbers often varied widely. Discrepancies in population numbers led to confusion, arguments and the inability to properly understand the true conservation status of the species.

While things were looking up for Asiatic lions in the 1930s, the decade that followed, defined by political turmoil and unrest, saw the promising conservation work to save the lions of the Gir Forest by the nawab and the SPFE thwarted.

As India moved towards independence from the British Empire in 1947, the country was partitioned into two new nation states, the Dominion of India (becoming known as the Republic of India in 1950) and Pakistan, a new country. As part of this process, Mahabat Khanji III had to decide which country Junagadh would accede to.

By 15 August 1947, the nawab had made his decision, announcing publicly that Junagadh would join Pakistan. Many regarded his choice as a disaster. Although the nawab was a Muslim, the vast majority of his citizens were Hindu. For them, acceding to Muslim Pakistan was fundamentally unacceptable. Their hearts, minds and future were with the Dominion of India, a largely Hindu country.

As unrest grew, the prince's administration began to falter. By October, his position was untenable. His only way out was to escape to Pakistan, leaving Junagadh to join the Dominion of India. Abandoning his ancestral home, his

forest and his lions was a terrible moment for Mahabat Khanji
III. On 24 October 1947, as the nawab boarded his flight to
the capital of Pakistan, Karachi, it's reported that he looked
mournfully towards Gir and tearfully said in a rather high-
pitched voice to anyone who might be listening, 'Who will
protect my lions now?'

Mahabat Khanji III was right to be concerned about leaving
his lions. Without him, and with the British administration
gone too, the world's last Asiatic lions were left defenceless,
unprotected and extremely vulnerable. This was a time of
political upheaval and uncertainty. The focus was on
establishing India as a new independent nation, which for
now took priority over conservation concerns. In the absence
of the prince and the British, their rules and regulations
regarding the hunting of lions no longer applied. In effect, an
open season on lions was declared.

Hunting a lion in the Gir Forest no longer required an
official permit. Anyone, princes or local people, could shoot a
big cat with impunity. Unfortunately, there are no official
records that show exactly how many lions were shot and killed
during this period. However, Mark Alexander Wynter-Blyth,
a British naturalist based in India, wrote in the *Journal of Bombay
Natural History* in 1949 that the 'slaughter was exceptional', and
that seven lions were hunted and killed in the Jetpur area of the
forest. Bagging a trophy lion may not have been possible just a
few years back, but it most certainly was now.

Without a ban on hunting, the sound of gunshot once
again echoed through the trees of the Gir Forest. And it wasn't
just lions that hunters had their eyes on. Other big mammals,
especially large herbivores, including chinkara and blackbuck,
were also gunned down in what the influential wildlife
conservationist E. P. Gee, in his book *The Wild Life of India*,
called 'a wholesale killing of wildlife' and Divyabhanusinh
described as an 'unprecedented onslaught'.

Although conservation work was neglected during India's
first years of transition from imperial rule to independence,
once political stability started to emerge, the roar from the

forest was heard once again and the plight of the country's lions was brought back into focus.

In 1948, the administration of the newly formed Saurashtra state – comprising Junagadh and more than 200 other princely states – began to put steps in place to reduce the number of lions being shot in and around the Gir Forest.

Saurashtra's prime minister, Jawaharlal Nehru, received pleas from home and abroad to protect the world's last Asiatic lions, including the Gujarat Natural History Society of Ahmedabad and the SPFE, as well as the Duke of Bedfordshire. They were very concerned about the welfare of Gir's lions, insisting that free-for-all shooting must desist and that their habitat, the Gir Forest, be protected from further deforestation. In response, Nehru announced that he was on the side of the lions. 'I have long been interested in the preservation of lions in India,' he said, 'and it would be a great pity if they were shot or otherwise allowed to suffer extinction'.

This 'great pity' for the plight of India's lions manifested itself in a number of practical and positive ways that were managed effectively, providing protection for the forest's lions. Although the region's royalty were still permitted to take part in lion *shikar*, the number of lions allowed to be killed was limited and strictly controlled by the then head of state Jam Saheb Digvijaysinhji, the Maharaja of Nawanagar. This was the first time that one authority was able to rigorously monitor lion hunting in all of the state's principalities – something that Mahabat Khanji III and his father Rasul Khanji had long striven for.

Official protection was also provided to the Gir Forest itself, with staff recruited to prevent the forest being illegally felled and logged. In February 1948, conservationist S. R. Pandya was recruited to protect the forest, although primarily to safeguard its production of timber, charcoal and firewood as a source of state income rather than to preserve lion habitat.

Pandya was one of India's first conservationists to recognise the link between the grazing of livestock in the forest by Maldhari and deforestation. He initiated a small programme to relocate twenty Maldhari villages to designated areas outside of the Gir Forest, where they could raise their livestock without affecting the income generated by the forest.

Another advocate for relocating the Maldhari, but for quite different reasons, was the visionary conservationist and naturalist K. S. Dharmakumarsinhji, popularly known as Dharma Bapa. A former game hunter, Bapa was dedicated to preserving the Asiatic lion as well as the Indian cheetah and the greater one-horned rhino, and published numerous papers suggesting new conservation strategies to protect India's rare species.

While working in the Gir Forest in 1948, Bapa noted that far more lions were slaughtered by people in response to their animals being killed by the big cats than in self-defence. Rather than just offer pastoralists compensation for their losses and hope that the lions would then be left alone, Bapa argued that it would be far better if all the Maldhari and their livestock were given new land on the outskirts of the forest. This way, conflict between people and Gir's lions could be practically eliminated and the lives of many lions saved.

Ahead of his time, Bapa urged that the forest be designated a national park, with some areas set aside for commercial forestry, but that the rest receive official protection, a place where wildlife can roam freely and safely, without fear of hunters or poachers. In *Gir Forest and the Saga of the Asiatic Lion*, the Indian naturalist and writer Sudipta Mitra describes Bapa's holistic vision for Gir as a 'forest where the roar of the lions will reverberate and will be enjoyed equally as the songs of the birds'.

Although just a relatively small number of Maldhari were moved at first, this initiative – combined with the tight control on lion hunting – started to show positive results. Official lion surveys in 1950 and 1955 showed that Asiatic lion numbers were on the up. The 1950 census recorded 227 lions and the 1955 survey showed an increase of 63 lions, making 290 in total.

These findings led Wynter-Blyth to declare that Gir's lions 'were in no danger of extinction, and an increase in numbers may well occur. Their only enemy seems to be man, and if left alone by him, I believe they will be able to look after themselves.'

As Saurashtra merged with Bombay State in 1956, splitting in 1960 to become the present-day state of Gujarat, princely *shikar* finally came to a complete end. Despite this, lion numbers began to fall. Wynter-Blyth's confidence proved to be misplaced.

Although Gir's lions were no longer hunted, they now faced new pressures created by the region's rising human population. Growing industrialisation, alongside a growth in cattle farming, turned prime lion habitat into land unfit for herbivores, making it much harder for lions to find prey.

When the Gujarat Forest Department (GFD) conducted its first Asiatic lion census in 1963 it counted 285 lions, just 5 fewer than the last count of 290 in 1955. But by the time of the next census in 1968, only 177 lions were recorded in and around the Gir Forest. In just five years, more than a hundred lions had been lost.

Alarmed by this rapid decline, the GFD responded quickly. A significant amount of land, where most of the lions roamed, was given state protection in 1965. Renamed the Gir Wildlife Sanctuary, this part of the forest became the first protected area in the Gujarat region. All flora and fauna in the sanctuary were now officially out of bounds for any further encroachment or development that might affect the lions' habitat or welfare.

In addition, the GFD also developed a land management plan that focused on restoring the forest's indigenous fauna, most of which had all but gone as a result of overgrazing by livestock. To give the land a chance to recover and revert to its natural state, a new resettlement programme to relocate Maldhari to new villages outside of the sanctuary was started in 1972.

As the Maldhari moved, taking their buffaloes, cows and goats with them, the native grasses and shrubs that had once covered the land began to grow back. The return of indigenous plant life was soon followed by the arrival of wild herbivores that had occupied the land before Maldhari livestock had grazed away the flora they had survived on.

Antelopes and deer, including chitals, sambars and nilgais, began to be seen again. Wild boars were back too, munching and crunching on the forest's luscious new greens. As herbivores found new food sources, they became food themselves, perfect prey for Gir's hungry lions that had been dining on Maldhari livestock in their absence.

As native prey became more abundant, the diet of Gir's lions changed markedly. Before the resettlement of the Maldhari, analysis of lion scat in the early 1970s showed that 75 per cent of the lions' diet consisted of livestock, mainly cattle and buffaloes. By 2010, though, with Maldhari and their animals gone, scat analysis showed that almost three-quarters of their diet was now wild prey. The GFD's plan to restore flora to bring back fauna had worked.

Conservationist H. S. Singh in his 2011 research paper 'A Conservation Success Story in the Otherwise Dire Megafauna Extinction Crisis: The Asiatic Lion (*Panthera leo persica*) of Gir Forest' noted that the number of native prey in protected areas 'increased by 10-fold between 1970 and 2010'. And that this abundance of natural prey was responsible for supporting 'an increase in the lion population from 180 animals in 1974 to 411 animals in 2010'. The return and management of healthy populations of herbivores proved to be a major factor contributing to the recovery of Asiatic lions from their decline at the close of the 1960s.

By restoring and protecting lion habitat and with goodwill from local people, the lion population began to rise steadily. As both social and territorial animals, however, lions need space to thrive. If they're not able to find new territories, behaviour and pride structures can be negatively affected. To offset this, the GFD made the whole of the Gir Forest a national park in

1975 and created additional sanctuaries outside the area. These included the Pania Wildlife Sanctuary to the east in 1989, the Mitiyala Wildlife Sanctuary also to the east in 2004 and finally the Girnar Wildlife Sanctuary to the north in 2007.

Collectively these sanctuaries are known as the Gir Protected Area (GPA) and cover more than 1,000 square kilometres in total. Under the jurisdiction of the GFD, each area's habitat is managed carefully and patrolled by wildlife rangers to protect lions with territories there. Wildlife rangers, including a team of women known as the Lion Queens, are equipped with state-of-the-art kit and carry out regular protection and surveillance operations to safeguard their lions against possible poachers. Passionate about their work, Gir's rangers have ensured that no lions in the GPA have been lost to poachers since 2014.

Rangers also monitor the forest and sanctuaries for sick or injured lions and are able to administer veterinary care or get them to on-site centres for prompt, specialised and often life-saving treatment. For example, in late 2014 when the GFD suspected that lions in the Pania Sanctuary had developed fluorosis, which causes tooth and bone decay, they immediately tasked the Anand Veterinary Institute with investigating further and initiating a plan to resolve the issue.

The GFD's work restoring, protecting and extending the lions' habitat has been one of the fundamental reasons for the dramatic pick-up in numbers of Asiatic lions since their low point in the late 1960s.

Although lions benefitted from a restored and protected habitat, thousands of open pits, originally dug out by Maldhari for use as wells, continued to pose a threat to their safety. This was because, as environmental journalist John Platt explained in *Lion* magazine, 'the open wells are 20 to 30 metres deep and are often surrounded by vegetation, which hides them from view'.

Unable to see the camouflaged wells, lions simply fell down them. Most sustained fatal injuries on their descent or drowned once they hit water. The few that survived were left with

debilitating injuries. From 2001 to 2008, twenty-eight lions –
more than double the number of lions killed by other threats
combined – died this way. Although this number might not
sound huge, Platt noted in his article that 'with so few lions in
Gir to begin with, these deaths put the entire species at risk'.

Alarmed by the number of lions dying because of the open
wells, in 2006 Kishore Kotecha, founder of the Wildlife
Conservation Trust of India, started raising funds for the
GFD to build lion-proof walls around Gir's dangerous wells.
Keen to add political weight to his fundraising campaign,
Kotecha approached Gujarat state governor Mona Sheth for
support. Unaware of the issue, Sheth was shocked when she
was shown a picture of an injured lion, saying afterwards how
the image moved her and caused her pain to think that 'the
king of the jungle would live helplessly for the rest of his life'.

Moved by the fate of the fallen lion, Sheth brought
momentum to Kotecha's campaign, printing leaflets and
spreading the word to a larger audience that the world's last
remaining Asiatic lions were threatened by Gir's open wells.
The funds subsequently raised were effectively used by the
GFD. Walls strong enough to withstand earthquakes were
built around more than 450 wells. As a result, the number of
lions dying because of wells fell significantly. In 2010, just
one lion lost its life this way.

However, lions continue to fall down the many thousands
of wells outside of the GPA. As recently as 2016, two lions
were reported in the Indian press as having done this. In
January 2016, a lion cub was found in a well in a village near
the Gir Forest, and in May that year a four-year-old male fell
down a 30-metre dry well in a village in Amreli. Thankfully,
the GFD were called to both scenes; the two lions were
rescued and released back into the wild safe and sound.

It's not just GFD staff that are looking out for Gir's wandering
lions. Since being settled outside of the forest, the positive

shift in attitudes from the Maldhari towards the big cats has also been a significant factor in the latter's survival, and the subject of studies by Meena Venkataraman, an independent researcher studying Asiatic lion behaviour and ecology.

In 2010, Venkataraman spoke with more than 3,000 Maldhari and other local people living in 84 villages in and around the GPA to find out more about their attitudes towards the lions they shared the landscape with. Her findings were published in a paper called 'Managing Success: Asiatic Lion Conservation, Interface Problems and Peoples' Perceptions in the Gir Protected Area' in 2014.

Even though lions take their livestock, Venkataraman found that the overwhelming attitude of the Maldhari to Gir's big cats was not one of anger or resentment, but rather one of 'pride and love for the lions'. When local people talked about Gir's lions, Venkataraman reported that they were described fondly 'as part of the family' and as 'the jewel of the forest'. Some local people have taken second names such as Sinh and Savaj, Gujarati words for lion.

Venkataraman's work also revealed that many Maldhari were actually pleased that lions lived nearby, because the big cats ate the herbivores that fed on their crops: 'Lions do help with this menace ... and we don't mind having them around,' said a local. Likewise, some people felt that the presence of lions acted as a deterrent to local thieves that steal animals and crops. She also found that 98 per cent of the people she engaged with said they would pass on information about sick or distressed lions to the GFD so that the animals could be helped.

While 95 per cent of the people Venkataraman spoke with said they had good relations with the GFD, 66 per cent felt the process for passing on compensation for livestock killed by lions was not well managed. However, Venkataraman also noted that as a counterpoint, some of the people surveyed said they never claimed compensation since 'it was the lion's natural food and culturally unacceptable to claim money for the death of sacred animals that provide great value and service when alive'.

In *Monster of God*, Quammen also talks about the positive feeling that Maldhari have about lions. He recounts a conversation he had with a 56-year-old Maldhari man called Bapu. Talking about the lions, Bapu tells Quammen that if Gir belongs to anyone, it is to the lions, not the people, adding: 'If [the lions] can't stay here where will they go? We're the intruders … if the lions go, the forest goes and everything else goes.'

Another 2014 study, 'Living with Lions: The Economics of Coexistence in the Gir Forests, India', led by Kausik Banerjee, a researcher from the Wildlife Institute of India, also supported the view that Maldhari communities contributed to the survival of Gir's Asiatic lions because the big cats were not viewed as a threat.

Banerjee found that when lions had taken livestock, it tended to be 'mainly unproductive cattle (such as bulls, ailing calves, aged and dying cattle)', which Maldhari were not too concerned about. He also learned that the 'free grazing rights and the compensation at current rates were additional profits for Maldhari families living inside Gir'.

The study reported that the profit made was equivalent to a 'person's annual minimal wage for 213 man days'. In addition, Maldhari were also able to make free use of forest resources, including water, firewood and forest topsoil for adding to manure, as well as have access to job opportunities with the GFD.

Fair compensation for the loss of livestock and use of forest resources has further fostered benign feelings towards the lions, tempering any resentment or anger that might be roused when cattle are lost and livelihoods threatened.

Positive attitudes were also shown after six lions, in separate incidents, were run over by goods trains in the early part of 2014. Their deaths were widely reported by local and national media, causing an outcry for lions to be better protected from trains. As a result, the railway authorities agreed to reduce the speed of their goods trains.

They also agreed to sensitise their drivers about the fragile status of Asiatic lions and make clear their important role in minimising the danger to lions posed by trains. The railway company then committed to building twenty-three underpasses for wild animals to use and the GFD put up barbed wire fencing around 28 kilometres of train track to help keep lions out of danger.

In November 2016, when one of Gir's most well-known lions, a fifteen-year-old male called Ram, was found dead in the forest, his body was taken away by officials to confirm the cause of death. Ratan Nala, deputy conservator of forests, told the BBC: 'We carried out a post mortem on him to ascertain the cause of death. He died from natural causes. He was cremated in the presence of government and local village officials.'

Such was the love and respect felt for Ram, described as 'beautiful' and 'flamboyant', that local people and forest officials fasted for a day to mourn his death.

Writing about the future of the lions in *Current Conservation* magazine in 2014, Venkataraman wrote that her experiences with the Maldhari convinced her that it was the love of lions from local people that helped keep them from extinction.

Community education and sensitisation projects have also created and reinforced feelings of positivity, rather than animosity, towards Gir's lions. So, too, have programmes that actively involve local young people in the day-to-day protection of the big cats.

Working as Vanya Prani Mitra (friends of wildlife), thousands of young people have earned much-needed income looking out for lions. For example, when lions enter a village, the Vanya Prani Mitra contact the GFD to come and take the lions away. While waiting for the GFD, they minimise the contact between villagers and the visiting lions, helping to keep the big cats safe.

As lion numbers continued to grow, they started moving outside of their new sanctuaries. Records of pugmarks, prey kills and details of sightings have shown lions to be present in more than 1,000 villages located outside of the GPA. It's estimated that today, 40 per cent of lions live in unprotected areas in Gujarat.

Since 2010, the GFD's Asiatic lion censuses have surveyed four new areas,* in addition to the Gir National Park and Sanctuary and three other sanctuaries† located inside the GPA, covering around 20,921 square kilometres. This has enabled the GFD to include lions outside of the GPA in their census counts.

Moving in and around villages home to more than 100,000 people, the unprotected landscapes that lions pass through are very different from their rich natural habitats in and around the Gir Forest. In *Current Conservation*, Venkataraman describes these areas as 'agro-pastoral ... industrial townships and highly populated villages'. Outside the GPA, there are few wild places, which means fewer wild herbivores for lions to prey on. As a consequence, and as Venkataraman points out, 'livestock depredation and incidences of lion attacks on people ... [have] increased'.

Inevitably, once-rare incidents of conflict between people and lions have started to become more widespread. In the last three years, fourteen people living outside of Gir's protected areas have been killed by lions. Six of these deaths happened in the first five months of 2016.

Three of them – a woman aged fifty-four, a sixty-year-old man and a boy of fourteen – were widely reported in the Indian media because they had also been eaten by the lions

---

* The four areas are: 1) the south-western coast (Sutrapada-Kodinar-Una-Veraval), 2) the south-eastern coast (Rajula-Jafrabad-Nageshri), 3) Savarkundla, Liliya and adjoining areas of Amreli and 4) Bhavnagar district.
† 1) the Girnar Sanctuary, 2) the Mitiyala Sanctuary and 3) the Pania Sanctuary.

that killed them. In response to these deaths, the GFD captured a pride of lions, some seventeen strong, known to roam the area where the three people were killed.

Once rounded up, the big cats were kept in cages for twenty-five days while their faeces were examined for human remains. Final analysis showed that one adult male and two young females had eaten people. The fourteen lions that showed no evidence of being man-eaters were released deep in the forest, far from where they were caught. The 'guilty' adult male was taken to Sakkarbaug Zoo in north Gujarat and the two lionesses confined to GFD rescue centres. All three lions will live the rest of their lives in captivity.

Portrayed by the press like a trial, this was the first time that Gir's lions had ever been captured, tested for evidence of man-eating and then 'punished' with imprisonment. Reiterating this, Govind Patel, a former GFD chief wildlife warden, told the *Indian Express*: 'Gujarat has never faced such a situation when lions have killed people and allegedly eaten them. Lions and Maldhari have lived together inside Gir for centuries.'

Even though local people were alarmed and frightened by the thought of man-eating lions close by, many of them still felt benevolently towards the big cats – as Gopal B. Kateshiya, a journalist from the *Indian Express* who visited the area where the three people had been eaten, found out when villagers told him that lions were 'sensible' and would only eat a human if they were very hungry.

Kateshiya also learned that since lions had started visiting villages in the region in the late 1980s, farmers, aware that lions tended to hunt in the evenings, 'stopped working on their farm at night unless unavoidable'. 'I used to take my cows and bullocks for grazing at 4am. But now, it's too risky to venture out before daybreak,' one farmer told Kateshiya.

Others have built walls around their houses and smallholdings, some topping 3 metres tall, to keep lions out.

Gujarat officials have also done their bit to help keep people safe and minimise conflict. Public meetings have been held in villages, with demonstrations showing people how to safely scare off unwanted lions. Vehicles equipped with loudspeakers have informed 'villagers of the do's and don'ts of living in lion territory' and leaflets have been distributed suggesting people sleep indoors and not outside on the ground.

There have also been promises to build sheds and provide tents for labourers who don't have anywhere to sleep at night, and compensation is paid to farmers who have lost livestock to lions. The GFD also continues to round up and relocate lions reported to be prowling in populated areas.

In 2014, moving lions accounted for 500 of the GFD's operations outside of the GPA. Regarding the scale of the GFD's work in reducing human–lion conflict, Gir Forest official Anshuman Sharma told the BBC that 'wildlife management has now become more of human management', and that staff worked all hours to 'ensure human life doesn't get affected as it's only due to the local people here that the lion population has survived'.

Fortunately, the GFD have not had to work alone regarding this and other challenges, as I learn when I visit the Asiatic lions at London Zoo.

Land of the Lions, London Zoo, London

The sign at the railway station says Sasan Gir. The station is small, with just one platform, and connects Sasan – a small town located at the edge of the Gir National Park and Wildlife Sanctuary – with the outside world. However, it's not a train that I'm waiting for – it's Asiatic lions. Because the train line runs through part of the park, some of the big cats can be spotted on the tracks beside the station.

I take a seat on the platform's wooden bench and wait a little longer before deciding to look elsewhere for lions. Perhaps it's too hot for them here out in the searing sunshine.

At more than 30 degrees, I know I'd rather be resting in a shady spot in this heat. As I leave the station, I pass through a small street full of colour and light. On a wall there's a mural of male peacocks, their emerald feathers extended like opulently painted fans.

There are a handful of market stalls and shops selling silky fabric for saris and spices for cooking, as well as a barber's and a tourist information office. Inside, the shops are just as colourful as the outside. Fixtures and fittings are painted in extravagant hues of fuchsia, turquoise and tangerine. All around, the air is filled with the faint aroma of incense.

It's only when I see a group of British nursery school children and their teachers chatting excitedly that I remember I'm not in Sasan. I'm not even in India. I'm in London Zoo, at the Land of the Lions exhibit that is home to four Asiatic lions.

The exhibit, around two years old, has been specifically designed to recreate Sasan in a London setting. Everything has been sourced from India and carefully curated – from the wooden bench I sat on at the station to pieces of signage and props in the shops – so it looks just like Sasan.

The exhibit has been made to resemble Sasan to help visitors appreciate just how closely people in and around the Gir landscape live to lions. It also helps to raise awareness that lions don't just live in Africa. Before the exhibit opened, few visitors appreciated that the Asiatic lions they were admiring came from India. Most assumed lions only live in Africa. Now, the zoo says, nine out of ten people who have visited Land of the Lions are aware that they've seen lions from India rather than Africa.

From the street area, I move on to a raised walkway designed to look like the rocky Girnar Hills, a site of holy pilgrimage for devout Hindus. From the top of the walkway, I look down onto a stretch of land dotted with trees and small shrubs. Laid out in the shade of a couple of trees are three lionesses, all of them sisters. The two youngest are seven years old and the eldest is nine. All three were born here at the zoo.

The trio are barely moving. As they doze, their rich honey fur picks up shafts of light that slice through the gaps in the trees. Just the occasional twitch of a tail and the opening of an eye let me know they've not quite given in to full sisterly slumber.

Suddenly, from under a tree just a few metres away, something stirs in the shade. It's another lion, a seven-year-old male that arrived here last year from Assiniboine Park Zoo in Winnipeg, central Canada.

He rolls over onto his back, stretches out, hind legs akimbo. He looks incredibly muscular. Big, too – much larger than I expected an Asiatic lion to be. He's also hirsute, with a dense mane full of curls and waves. Again, I'm surprised, since many male Asiatic lions are described as being scantily maned. The nursery school children are watching him too, equally impressed with the size and stature of the Asian king of the jungle.

It's hoped that the lions here will mate. They're part of an international captive breeding programme to help ensure the genes of Asiatic lions survive should anything catastrophic happen to the wild lions at Gir. There are captive-bred Asiatic lions in Gujarat too. Around fifty reside at Sakkarbaug Zoo, and are considered to be a significant conservation breeding population.

At Sakkarbaug, the Zoological Society of London (ZSL) – the conservation wing of London Zoo – works with veterinarians and keepers, sharing best-practice wildlife health techniques and developing high standards of care. It also promotes techniques that help captive lions develop natural behaviours – such as hunting, foraging and scent marking – that enrich the lives of the lions at Sakkarbaug by keeping them engaged and active.

It's not just captive breeding that ZSL is involved with at Gujarat. A partnership between ZSL, the GFD and the Wildlife Institute of India involves plenty of on-the-ground support.

Experts from ZSL have travelled to India and shared their conservation knowledge with wildlife professionals working at Gir. Up until October 2017, ZSL provided training support and worked closely with frontline staff to streamline patrol techniques as well as training in state-of-the-art equipment used for the rescue and rehabilitation of lions that may be at risk of injury.

Supporting the excellent community work already carried out by GFD, ZSL has also worked with Sakkarbaug Zoo designing its education masterplan and communicating information publically so that people and lions can continue to live together conflict-free.

I take a last look at the lions. All four are still motionless, stretched out and relaxed in the shade, as if they are on holiday. They're oblivious to the delight of the small children staring down on them in their oasis of calm in the heart of one of the world's busiest capital cities. I'm followed out from the walkway by the children and their teachers, as if we've all collectively agreed that it's best to let sleeping lions be.

Outside of the GPA, Asiatic lions face many other threats besides people. There are the open wells, at least 1,900 of them; most without lion-proof walls around them. Electric fences are also deadly for lions. Barbed wire can kill, too. In June 2016, the *Economic Times* reported that a lioness died after getting trapped in barbed wire that had been put up around a farmhouse. Lions have even fallen into the sea.

In January 2016, UK newspaper the *Daily Express* showed pictures of the dramatic rescue of a male lion that had wandered 29 or so kilometres from the Gir Wildlife Sanctuary to the heavily populated and bustling Jafrabad Port on the Arabian Sea. Fishermen spotted the lion walking around rocks and called the GFD, who arrived just in time to see the lion plunge into the sea. Swiftly darted and tranquillised, the 180-kilogram big cat was then lifted from the water and taken to a veterinary centre to be checked over. He was later released back into the Gir Forest.

Lions are also hit and killed by cars and trains as they cross roads and wander over rail tracks. Since lions have started moving into the human landscape, more than 260 lions have lost their lives in the last 5 years because of accidents like these.

Today, lions living outside of the Gir Forest and the surrounding areas face similar threats that lions living in the forest once faced, before those habitats were officially protected. Not wanting history to repeat itself and lion numbers to start declining again, in May 2015 the Gujarat government set up a high-level task force dedicated to the safety of lions outside of protected areas.

Scientists and conservationists attached to the task force were charged with creating dispersal corridors between protected and unprotected land to enable lions to travel safely between the two without having to risk encounters with people, traffic, trains and wells. They are also considering creating new protected habitats to help keep their lion overspill safe.

Posting GFD wildlife rangers outside of the protected areas to safeguard the lions was also being considered, said P. K. Taneja, the then additional chief secretary of the GFD, speaking to the *Times of India*. As well as ensuring safer new habitats and moving wildlife rangers to unprotected areas, the management of human and lion conflict is also being stepped up. C. N. Pandey, the GFD's principal chief conservator of forests, also speaking to the *Times of India*, said there were 'measures for better conflict measurement' in place, including the establishment of 'regular communications with locals and farmers' and the provision of training to help forest staff deal with conflict as it arises.

Alongside the task force's mission to find the most effective methods to make Gir's migrating lions less vulnerable, there were also calls to relocate some satellite populations elsewhere. As well as facing the dangers of open wells, traffic and trains, Gir's lions live relatively closely together and share interlinked habitats. If disease broke out, or there were a devastating fire or flooding, the world could quite easily lose its entire population of Asiatic lions in one fell swoop.

For example, many Maldhari and local people have dogs that carry Canine Distemper Virus (CDV), which Asiatic lions, just like their cousins in Africa, can catch with fatal consequences. Regarding the seriousness of this threat, a senior conservation biologist at the Wildlife Institute of India (WII), Ravi Chellam, spoke with the BBC in 2014 noting that the CDV epidemic in Tanzania in 1994 took out more than 1,000 lions in just a few days and that 'something of that sort can't be ruled out ... as a sizeable population lives outside the forest and is exposed to infection'.

In 1990, the WII advised that such an outcome could be avoided by relocating some of Gujarat's lions to another wildlife sanctuary with the right habitat and prey base where a new population of lions could be established, thereby safeguarding the species. Two years later, in 1992, the Indian government and WII set up a lion reintroduction project to do just this.

One of the project's first tasks was to find a site for a possible reintroduction. Working alongside officials from state forest departments around India, scientists and conservationists drew up a shortlist of five locations that might be the perfect des res for lions relocated from Gir.

Three of the sites, the Sita Mata Wildlife Sanctuary, the Darrah–Jawahar Sagar Wildlife Sanctuary and the Kumbhalgarh Wildlife Sanctuary, were located in Rajasthan, India's largest state, in the west of the country. The Kuno Wildlife Sanctuary, around 1,126 kilometres from Gir in the state of Madhya Pradesh in central India, was also shortlisted, as was the Barda Wildlife Sanctuary, fewer than 112 kilometres away from the lions' existing home.

After site visits and detailed consideration of each sanctuary's habitat, prey base, water sources and levels of human encroachment, the Kuno Wildlife Sanctuary in Madhya Pradesh was selected as the area best suited for rehoming some of Gujarat's lions. Led by Ravi Chellam, a detailed and robust framework was then established to kick-start the move. It featured key activities and timelines to enable a successful translocation of lions to Madhya Pradesh by 2005.

This included improving the habitat that the lions would occupy at Kuno Wildlife Sanctuary, as well as relocating twenty-four villages outside of the sanctuary and away from the lions to reduce the likelihood of human–lion conflict. By 2003, all 24 villages and the 1,500 or so families that lived in them were resettled outside of the land set aside for Gir's lions. Habitat was also restored, complete with a robust prey base.

When 2004 arrived, the Kuno Wildlife Sanctuary was ready and set to receive nineteen Asiatic lions. However, Kuno staff were in for a long wait. The government of Gujarat, which had never actually formally agreed to share its lions, flatly refused to part with its big cats.

Although government and conservation officials in Madhya Pradesh may have been happy with the suitability of Kuno as the lions' second home, many conservationists and state representatives in Gujarat were not. Critics of the proposed move pointed out that although Madhya Pradesh might be known as the Tiger State, it had a shockingly poor protection record where the endangered Bengal tiger was concerned. They noted that from 2002 to 2012, 453 tigers from 6 of Madhya Pradesh's tiger reserves had been killed, mainly by poachers.

In a decade, the state had lost almost half the number of tigers killed globally. And to make matters worse, Madhya Pradesh appeared to be doing little to ensure that tiger poachers were caught, prosecuted and imprisoned. According to The Hindu, only 'two cases of tiger poaching were punished' after a number of tigers had been killed.

The state's conservation record was further tarnished when the field director of Madhya Pradesh's Panna Tiger Reserve alleged that forest officials had colluded with poachers and posed a significant threat to the safety of the tigers at the reserve. If Madhya Pradesh wasn't able to protect its own precious tigers from poachers, what guarantee was there for the safety of Gir's lions, reasoned concerned conservationists in Gujarat?

When in May 2012 the Madhya Pradesh tourism office posted images of Asiatic lions on its website, even though there were as yet no lions in the region, relations between the two states became further strained. Opponents of the move said that Madhya Pradesh was more interested in the tourist money that Asiatic lions might bring in than protecting the species. Even Kishore Kotecha, who initially supported the translocation, was driven to wonder if Madhya Pradesh was more interested in having lions for tourism rather than conservation.

Other Gujarat conservationists, including the then state forest minister Mangubhai Patel, also pointed out that since the emergence of the translocation plans in the early 1990s, lions at Gir had been doing well where they were. Population numbers were increasing and looked set to continue this way. Lions were also establishing new territories outside of the forest, which with protection from the GFD were now safe, enabling new prides to thrive.

Patel maintained that since lions were establishing new satellite populations in separate territories, the concern that the world could lose an entire species because of disease or natural disaster no longer applied. 'There is no need to shift lions from Gir. We will ensure their survival here,' Patel confidently told the *Times of India*.

Reference was also made to Gujarat's initiative known as the 'Long Term Conservation of Asiatic Lion Plan', which had been lobbying the state for funding since 2009 to help make Gir's habitat safer for lions by building a substantial new ring road to divert traffic away from lion territories inside the

Gir Forest and surrounding areas, including Pania, Mitiyala and Girnar.

One hundred and seventy-one miles long, featuring sixteen underpasses and fourteen flyovers, the ring road project would, its supporters said, connect villages around the edges of the protected areas so that they no longer needed to travel through them to do so. Pradeep Khanna, principal chief conservator of forests in Gujarat, told the *Times of India* that with the ring road in place, only vehicles that 'belong to villages inside the sanctuary will be permitted ... This would be safe for animals and reduce the risk of poaching.'

Khanna maintained that restricting traffic in lion country would avoid accidents, referring to an incident where 'a lion died after it fell off a bridge because it was blinded by the headlight of a speeding vehicle'. A ring road would also deter poachers because they would no longer be able to enter the forest in vehicles to make quick getaways with bodies of illegally slain lions.

As well as the construction of a huge ring road, Gujarat's Long Term Conservation of Asiatic Lion Plan also applied for funding to support the relocation of more Maldhari outside the forest area and for further improvements to lion habitats, as well as funds for better lion rescue facilities, more education initiatives to reduce human–lion conflict and eco-tourism development.

Local people were also reluctant to see their lions go. In *Gir Forest and the Saga of the Asiatic Lion*, Mitra says that a 'strong wave of protection against translocation from Gir to any other state arose from the people'. Considered 'family members', it was thought fundamentally wrong to remove the big cats from Gujarat and its people.

This was a sentiment upheld by Gujarat state officials, including a Jassu Barad, who declared: 'The lion is part of [our] folklore. It just cannot settle in a new environment.' An unnamed GFD officer also told the *Times of India*: 'Unlike people [in Gujarat] who don't mind the presence of lions, the people in [Madhya Pradesh] feel threatened and even pose a danger to the lions.'

In response to Gujarat's ongoing refusal to release any of its lions for translocation to Kuno, WWF–India, who supported the move, filed a petition with the Supreme Court of India in 1995. The petition requested that Gujarat be legally bound to hand over some of its lions for reintroduction purposes to Madhya Pradesh.

In April 2013, eighteen years after the petition had been filed, the Supreme Court finally delivered a judgement that Gujarat must transfer some of its lions to Madhya Pradesh by October. This meant that the state was now legally obliged to hand over some of its big cats and assist the Kuno Wildlife Sanctuary with its plans to introduce Asiatic lions to the region.

The ruling did not go down well with either local people or Gujarat officials. The news was taken so badly that volunteers from one of the region's nature clubs staged a protest march, threatening mass suicide if the ruling from the Supreme Court was not overturned. And, according to Craig Packer, director of the Lion Research Center at the University of Minnesota in the United States, in his book *Lions in the Balance: Man-Eaters, Manes, and Men with Guns*, 'distraught Gujaratis went on a hunger strike protesting the decision; a dozen people died'.

Gujarat officials took a more pragmatic approach, but remained steadfast in their determination to keep all of their lions in-state. By July, they filed for a review of the Supreme Court's ruling on the destiny of its lions. Their reasons for petitioning a review of the ruling were plentiful and detailed.

For example, tigers were a worry. Gujarat conservationists were concerned that cubs had been recently sighted at Kuno, indicating the presence of tigers. They pointed out that because lions and tigers are both territorial and lions are known to generally come off worse in confrontations, putting Gir's rare lions at risk of encountering tigers was reckless.

A more pressing concern, though, was the effect that breaking up groups of lions might have. Removing members of a pride can affect hunting and communal cub raising, creating a disturbance and breakdown of a pride's structure, opponents of the move argued. Lions were far more likely to thrive if they remained local, the Gujarat petition maintained – requesting that a second home be piloted in a region with the same climate as Gujarat and where their progress could be monitored closely and learned from.

They maintained that the Bhavnagar Amreli Forest in the district of Amreli, east of the Gir National Park, would make a much better second home than the Kuno Wildlife Sanctuary. It had a topography and dry climate very similar to the Gir Forest, unlike the wetter climate of Kuno. And it also had a good prey base, as well as close access to GFD wildlife teams experienced in protecting and preserving lions. Another advantage, the Gujarat petition pointed out, was that a pilot at Amreli would also be 'in tune with international guidelines', which Kuno would not be.

The guidelines referred to were from the IUCN, issued in July 2013. Petitioners pointed out that the IUCN guidelines recommended 'a natural corridor between the original site and the new site', and that only as a last resort should a site without these prerequisites be considered, 'when no opportunities for the reintroduction into the original site [exist]'. Kuno Wildlife Sanctuary didn't meet any of these recommendations. There was no natural corridor and Amreli had not been considered for reintroduction either.

The guidelines also stated that any proposed new site must assess the attitudes of local people towards the translocation, especially if one of the reasons the species went into decline was the result of over-hunting or habitat loss. In Madhya Pradesh, both these activities had contributed to the region's historic loss of lions.

And, as Gujarat's petitioners pointed out, the state was now home to an alarming and growing gun culture. Hunting and poaching in Madhya Pradesh was still a clear and present danger to lions. For example, in May 2013, the *Times of India* reported on the level of legal gun ownership in the Sheopur district, where the Kuno Wildlife Sanctuary is based. The newspaper stated that in Sheopur there were an estimated 4,800 firearms for a population of 600,000, and that the number of unlicensed weapons was likely to be much higher as this was the tip of Chambal valley, which was also 'infested' with bandits.

Not only were there far fewer guns in Junagadh district, translocation opponents argued, but in areas around the Gir Sanctuary where lions roamed freely, the government of Gujarat had stopped issuing firearm licences altogether. No one living there could legally own a gun at all. This was a move unlikely to be repeated in Sheopur, the *Times of India* noted, because such an action would make the 'local populace resentful of the lions' presence'.

The high level of gun ownership in Sheopur even caused Ravi Chellam, the lead scientist working on the relocation plans, to write in one of his reports that poaching was a serious concern for conservationists, describing the situation as 'unacceptable and dangerous'.

In more dramatic fashion, the *Times of India* described Sheopur as a 'wild west' and 'safe haven for gun toters'. It also remarked that if lions were translocated from Gir to Kuno, they would find their former home 'a heaven compared to Sheopur where game hunting for meat is common'. Sending lions to Kuno 'is as good as sending them to a firing squad', the newspaper concluded.

The presence of hunters and their weapons in Sheopur had also been a concern for India's Ministry of the Forest and Environment when it vetoed the introduction of cheetahs from Africa to the Kuno Wildlife Sanctuary in around 2011. If the authorities thought poachers were too much of a threat to translocate cheetahs, then surely this verdict must also apply to lions, especially when any forester 'would tell you [it's] easier to kill a lion than a cheetah with a gun', reasoned conservationists who wanted lions to remain at Gir.

India's poor record in successfully translocating animals – not just lions, but other species too, including rhinos, elephants and leopards – was cited as another reason to leave

India's last remaining lions at Gir. In support of this claim, the Empower Foundation, an Indian NGO dedicated to reducing conflict between people and animals, produced a report called *Save the Lion, Say No to Translocation*.

After analysing ten Indian translocation projects, the report stated that only 16 per cent of introduced animals settled successfully in their new habitat. The remaining animals, 84 per cent of them, either died, were 'killed by locals due to severe conflicts', or were returned to their original habitat. Such dire statistics left the report's authors to conclude that 'India has had no majorly successful translocations'. The report also flagged up that India's record of previously translocating lions was especially concerning.

Back in 1904, three lion cubs from Africa had been released at Kuno. The project was a failure. All three were shot and killed. Later on, in 1956, three Asiatic lions from Gir were translocated to the Chandra Prabha Wildlife Sanctuary in Uttar Pradesh in northern India. They also died – this time as a result of 'inadequate area, lack of systematic monitoring and unrestricted movement of grazing animals'.

Undeterred by these failures, four decades later, between 1997 and 2001, twenty-two Asiatic lions were translocated to various different sites around India. All twenty-two lions were returned. The threats to their welfare from local people had been considered too much of a risk to keep the big cats *in situ*.

All translocations cause animals chronic stress, Jalpesh Mehta, team leader at the Empower Foundation told the *Times of India*, making them vulnerable in their new environments and likely to succumb to disease or predation. Moving lions to Kuno was against 'their overall interest', he told the newspaper, ultimately causing them to 'become victims of chronic stress, disease, reproductive issues and predation'.

On 14 August 2014, the Gujarat government's fight to stop the translocation of its wild Asiatic lions appeared to be over.

Their request for a review of the order to send Gir's lions to Madhya Pradesh was dismissed by the Supreme Court.

India had at last spoken and ruled on the matter. Gujarat must hand over some of its Asiatic lions to the Kuno Wildlife Sanctuary within six months. Representatives at Kuno felt confident, after decades, that finally some lions would be with them soon. So confident, in fact, that Kuno's retired chief wildlife warden, H. S. Pabla, told the *Times of India* in June 2013 that everything was now on track and 'there should not be any other problem in translocation'. Pabla, however, was wrong. Despite the state deadline, official reminders and urgent notifications from the government of Madhya Pradesh, no lions left Gujarat state. Gir's lions, it seemed, were not for moving.

Concerned by Gujarat's stalling and facing the real possibility that a translocation from Gir still might not take place, the Madhya Pradesh government started to look elsewhere for its lions. Its first port of call was Hyderabad Zoo (Nehru Zoological Park), based at Telangana in southern India, which ran an Asiatic lion captive breeding programme. Although Hyderabad Zoo was at first prepared to share some of its lions with Kuno, recent translocations of its captive-bred lions into the wild had not gone well.

In April 2013, two Asiatic lions (a female called Lakshmi and a male called Vishnu) were moved from Hyderabad Zoo to Kanpur Zoo (Allen Forest Zoo) in Uttar Pradesh, before being moved in September 2014 to Etawah Wildlife Safari Park, also in Uttar Pradesh. By the close of October, Lakshmi was dead. Before she died, Lakshmi had lost weight and had become paralysed. Vishnu displayed similar symptoms and was kept under close observation by a number of veterinary experts who noted that because of paralysis in his back legs, the lion was unable to walk properly.

Within a fortnight, Vishnu was dead, too. Autopsies on the pair revealed that their internal organs had failed, causing cardio-respiratory failure and death. The failure of their organs was attributed to a mystery virus. Possibly nervous about the bad publicity that might be generated if more of its lions died, Hyderabad Zoo never shared any of its lions with Kuno, leaving Madhya Pradesh without lions.

Then in April 2015 a new development, instigated by the government, further dashed Madhya Pradesh's hopes of receiving any Asiatic lions. It unexpectedly announced that the Kuno Wildlife Sanctuary, approximately 217 square kilometres in size, was simply not large enough to be a safe second home for translocated lions. To accommodate Gir's lions, Kuno would need to double in size, to cover at least 434 square kilometres, government officials insisted.

Meeting this new criterion was problematic for Kuno. To extend its boundaries, it would need to relocate several large villages which would be expensive as well as unpopular with the people who lived there. Forest Department officials from Kuno stated that this was also unnecessary because the sanctuary, if you included its buffer zone of more than 643 square kilometres, was already large enough – in fact, 209 square kilometres larger than requested – for Gir's lions to establish themselves.

However, rather than integrate the buffer zone into the sanctuary to meet the government's criteria, Madhya Pradesh officials did nothing. A report in the *Times of India* said that the state was 'dragging its feet' because a consequence of the integration would mean that up to 10 square kilometres of former buffer zone would then have to be 'reserved as an eco-sensitive zone' where no commercial, building or industrial development could take place.

The government's new concerns weren't just limited to the size of the Kuno Wildlife Sanctuary, though. It also wanted Kuno to carry out an extensive study of its prey base. The survey would need to meet agreed scientific standards and be conducted over the course of a year, to take account of the impact of seasonal climatic changes on both flora and fauna.

Two months later in June 2015, the state Ministry of Environment, Forests and Climate Change stated that it would be filing an affidavit with the Supreme Court, confirming concerns raised by two petitions lodged earlier that year by the Gujarat State Board for Wildlife and the Wildlife Conservation Trust about moving lions from Gir to the Kuno Wildlife Sanctuary.

An anonymous senior environment ministry official told *Mint*, India's second-largest business newspaper, that moving lions from Gir would be 'detrimental to their breeding and survival' because of their tight pride and group structures. He also told the newspaper that after a meeting of wildlife experts chaired by the state environment minister Prakash Javadekar, it was agreed that moving lions to Kuno 'could destabilise their population'.

The tiger corridor between Ranthambore National Park in Rajasthan and Kuno was also an issue. Some experts were concerned that lions would not be able to coexist safely with tigers because of territorial fighting. Also, what might happen if lions used the corridor and left Kuno, setting up new territories at Ranthambore? How would lions moving into the park at Ranthambore fare? To answer these questions, the Rajasthan government and the National Tiger Conservation Authority were asked to get involved and share their preparedness plans should lions move upstate.

Of course, the government's intention to file the affidavit did not go down well with conservationists in favour of moving Gir's lions. Among them was Fayaz Khudsar, a biologist at Delhi University, who had been instrumental in the Supreme Court's decision to overrule Gujarat's call to stop the translocations of its lions to Madhya Pradesh. He told the *Hindustan Times* that it was unlikely that any tigers from the Ranthambore National Park would be a threat to introduced lions, adding that, in fact, a paper published by Mahesh Rangarajan, a wildlife historian, stated that before lions in the region were all hunted down, lions and tigers in Shivpuri district (in Madhya Pradesh) had coexisted together just fine.

In the *Times of India*, Ajey Dubey, secretary of Prayatna, an environmental action group at the 'forefront of the campaign to move lions to Madhya Pradesh', was reported as saying 'We will fight tooth and nail if they file any such affidavit,' maintaining that the original plans to introduce just six lions from Gujarat to the Kuno Wildlife Sanctuary were scientifically robust and would have no impact on the pride structure of the lions that remained at Gir.

Dubey went on to say that 'the real threat to the king of the jungle' was in Gujarat where after recent heavy rains, nine lions drowned in the Shetrunji River. Dubey added that Prayatna was preparing to file new documents 'which will highlight that forest areas in Gujarat are now saturated and that the population growth is leading to lion–human conflicts'.

As the close of 2015 approached, the likelihood of Madhya Pradesh housing a population of Asiatic lions seemed further away than ever, especially when Prakash Javadekar, the state minister for the environment, informed members of the Indian parliament that any translocation of lions from Gujarat would take at least twenty-five years.

'The whole programme encompasses action for over 25 years,' Javadekar wrote in his letter to parliament, adding that the timescale was reached 'after consultation with experts, who were of the view that the translocation of lions was not immediately possible'.

Madhya Pradesh's response to this news was not positive. In February 2016, the Forest Department posted a statement on its website. The statement said that everything that could be done to prepare the Kuno Wildlife Sanctuary for Gir's lions had been done, including spending almost £650,000/ US$839,614 improving and extending the habitat, as well as relocating 1,545 families. The website post went on to say: 'If the stalemate is not resolved expeditiously, the state may have to consider introducing tigers in the sanctuary as wasting such a prime habitat is not in the interest of conservation of wildlife in the state'.

The idea of homing tigers instead of lions at the sanctuary had been suggested after a team from the National Tiger

Conservation Authority had visited the area as part of a study, Narendra Kumar, Kuno's principal chief conservator of forests, told the *Hindustan Times*. Kumar also told the newspaper that Kuno simply could not wait twenty-five years for lions: 'We can't wait for such a long time. The habitat has been readied for a carnivore and we will have to do something there.'

While Madhya Pradesh announced its impatience with the new scheduling for translocating Asiatic lions from Gir, the government of Gujarat made unexpected news of its own. On 24 February 2016, the *Times of India* reported that Gujarat had officially dropped its proposal for state funding via its Long Term Conservation of Asiatic Lion Plan. Why, after years of campaigning and lobbying for state support to deliver these initiatives to further safeguard its lions, the Gujarat government withdrew its plans, to date remains undisclosed.

As it stands, Gujarat and Madhya Pradesh remain at loggerheads. Since 1992, when the reintroduction project was formally established to help protect Asiatic lions from extinction by creating a new population of the cats elsewhere in India, not a single lion has been translocated. As the lion population in Gujarat has grown, it's had little choice but to enter a human landscape described by Indian environmentalist Takhubhai Sansur as a 'death field for the Asiatic lion'.

Many conservationists feel that Gujarat's stalling is due to politics rather than the good of the species. In 2014, Ravi Chellam told the BBC that he believed the state was 'playing politics', and that politicians in the past had fought to keep the lions in the state as a vote winner because of the symbolic importance of the lion for many Gujarati people, as well as the tourist revenue that the lions attract. Wildlife activist Bhikhabhai Jethava also told the BBC that Gujarat's lions were 'exploited ... for votes'.

It's a stalemate that causes frustration, not just among conservationists in India but wider afield, too. In 2014, Luke Hunter, president of wild cat conservation organisation Panthera, made his feelings plain in *Scientific American* magazine, stating that while Gujarat should be commended for its conservation efforts, this expertise should be shared so 'that lions persist in India regardless of how they do in Gujarat'.

Hunter himself has had direct experience of successfully translocating more than 500 lions in 40 different populations in southern Africa, and told the magazine that he believed moving Asiatic lions in India, combined with the experience in handling lions that conservationists in Gujarat possess, would have a good chance of succeeding.

For now, India's lions seem set to remain in the Gir National Park and surrounding sanctuaries of Gujarat. It looks likely that the population will – with the continuing protection of the GFD and support from local people – continue to grow. Hopefully, in time, an additional site can be found where a new population of Asiatic lions can thrive, safeguarding the future of the species in India.

CHAPTER SIX

People Love Lions – Part II

> When we started here, we rarely saw lions and certainly
> never thought we'd observe them living harmoniously
> alongside people. Now lions have tripled in number and can
> live safely near local communities. They provide hope that
> lions can thrive throughout Africa.
>
> Stephanie Dolrenry, co-founder and
director of science at Lion Guardians

When Cecil the lion was killed in Zimbabwe by
American trophy hunter Walter Palmer in July 2015,
his death shone a global spotlight on the plight and vulnerability
of lions in Africa. Shocked and outraged by the way Cecil was
hunted down, countless people around the world held vigils,
joined protest marches and donated significant amounts of
money to fund lion conservation projects to help save Africa's
last lions.

Before Cecil went down, though, thousands of people in
Africa and beyond the continent had been working both on a
professional and voluntary basis to save *Panthera leo* from
extinction in the wild. Numerous wildlife charities,
foundations and organisations – funded by individuals,
institutions and government bodies – had established local
projects and initiatives to help protect lions from the current
biggest threats to their survival.

In East Africa, notably Kenya and Tanzania, conflict
between big cats and people, especially in remote regions
outside protected areas, is lethal for lions. At least one lion a
week is killed – either shot, speared or poisoned – for taking

livestock. For many farmers and pastoralists, killing lions is the most effective way to stop them from offending – a simple but deadly insurance to protect their livelihoods from the unforgiving jaws of hungry lions.

Of course, taking lions out of the equation by killing them isn't the only way that livestock can be kept safe. Lion protection projects, run and funded by conservation organisations and charities, have been helping traditional pastoral communities find other ways to safeguard their livestock and livelihoods against lion attacks.

With names like Lion Guardians, Lion Defenders and Wildlife Warriors, these conservation programmes and projects work with local communities to change attitudes and behaviour towards lions so they're no longer seen as a threat, but instead as creatures of value that deserve protection rather than eradication.

The Lion Guardians programme has been working with Maasai communities to save lions' lives in East Africa since January 2007. It was established in the communally owned Maasai lands around the Amboseli National Park in southern Kenya, where lions were once on the brink of extinction.

For hundreds of years, Maasai people lived in relative harmony with the lions they shared the land with. True, lions were still killed, mainly in traditional hunts known as *olamayio*, when older boys on the cusp of manhood killed lions to demonstrate their emerging warrior skills, and by pastoralists in revenge for the occasional death of a cow or goat. But both then had little impact on lion numbers.

By the 1950s, the situation had changed dramatically. Kenya was now home to one of the fastest-growing populations on earth. As the number of Maasai grew, so too did their need for livestock and grazing land. By 2013, more than 35,000 Maasai grazed more than 2 million head of livestock on land around the Amboseli National Park. Inevitably, as lion habitat began to shrink and prey numbers along with it, conflict between people and lions began to escalate. Lions began to take more and more livestock. People hated them for it and so killed them indiscriminately.

By 1993, Amboseli National Park had lost all its lions to Maasai spears or poison. Outside the park, between 2001 and 2005, eighty-eight lions died the same way. In 2006, another forty-two were killed. This was the year that Leela Hazzah and Stephanie Dolrenry, both conservation biologists working in a team with other scientists and local Maasai warriors, developed the Lion Guardians concept as a new way to significantly reduce conflict between local people and lions.

For years, the scientists lived and worked closely with the local Maasai community, sharing their everyday lives and gaining insights into their traditions. Immersed in the local culture, the team came to the conclusion that the two biggest threats presented to lions living near them were the warriors' *olamayio* hunts and the communities' retaliatory killings of big cats for taking livestock.

The team appreciated that much of the draw of lion killing was linked to the prestige and respect that lion killers, typically young men known as warriors, received from their community. And so, a plan based on an idea from the warriors was formulated to save lions by working with the local value system, but switching it around so that the men who once killed lions would – like bodyguards – protect them instead, and gain status as Lion Guardians, preserving lions rather than persecuting them.

'We don't try to change the mindset of the warriors to thinking that lion killing is "bad",' Hazzah told the Society for Conservation Biology, an American organisation dedicated to protecting the earth's biodiversity, in 2012: 'Instead we … utiliz[e] all of the positive Maasai cultural values towards lions to drive the conservation effort and provide them with an example that living with and protecting lions can provide them with similar prestige.'

One example of how the Lion Guardian programme adapts traditional Maasai customs to benefit rather than harm lions is shown in the practice of lion naming. Traditionally, during an *olamayio* when a young man became a warrior by killing his first lion, he was given a 'lion name' that became forever his and generated immense prestige. Now, when Lion Guardians track their lions, they give any cubs a special name and the animal becomes forever 'theirs' to protect instead.

In 2013, Kamunu Saitoti, a Lion Guardian at the Amboseli-Tsavo ecosystem, told the *Sydney Morning Herald* that a 'very deep connection' existed between Lion Guardians and named lions, a connection he liked as one that can 'only be compared to the bond between best friends, or the feelings you have for your best cows'. Now, when cubs are born, warriors who once delighted in killing lions celebrate because 'these lions aren't just lions to them anymore, they are individuals'.

An important aspect of a Lion Guardian's work is stopping lion hunts, both *olamayio* and retaliatory. Sometimes, Lion Guardians are able to simply persuade men not to take part in the hunts, primarily because of the standing and respect they have in their local communities. In this way, during 2010 to 2013 Lion Guardians prevented 133 lion hunts from taking place.

Working in their home communities and dressed in traditional clothing, Lion Guardians' roles are salaried and revered as positions of responsibility. Young Maasai men who work as Lion Guardians are able to protect and support their families, without the cyclical worries of drought and disease that cattle herders endure. It's said that you'll often hear a Lion Guardian say they 'love their lions like cows' or that they 'can milk lions' because lions are now seen as having value, providing livelihoods where few other employment opportunities exist. And for this, communities afford their Lion Guardians great respect.

Maasai men who want to become Lion Guardians don't need a formal education to be recruited. As part of their training, they learn to read and write, as well as use GPS and telemetry to track and monitor lions. The acquisition of these technical skills, combined with their traditional hunting prowess and local knowledge of the area, enables them to accurately track the movements and position of lions and then let local pastoralists know which areas to avoid herding their cattle in, reducing opportunities for conflict.

Tracking skills are used to find and bring home lost cattle and herders, as well as children who have strayed into the bush. Lion Guardians also help families reinforce their bomas (livestock enclosures) to make them lion-proof and keep their livestock safe, again reducing conflict between people and lions.

In her paper 'Conservation and Monitoring of a Persecuted African Lion Population by Maasai Warriors' published in *Conservation Biology* in 2014, Stephanie Dolrenry reported that by the end of the first year of the Lion Guardians programme, 98 per cent of Guardians 'could read and write their name, the name of their zone and area, the time, and numerals'.

The Lion Guardians' self-esteem was also increased, Dolrenry noted, quoting Guardians as saying: 'I am proud because the Lion Guardians programme has made me literate' and 'The programme has increased our status in the community because we are now literate. With our GPS and scientific forms, it has placed us in a different league.'

Dolrenry also noted that Lion Guardians also spoke of better coexistence between themselves and their communities and lions, citing the following quotes: 'The Lion Guardians program has brought peace between the Maasai and lions', 'We have a caring attitude towards livestock and lions, we act as a mitigating tool' and 'The project has made a previous enemy into a friend.'

Around nine years ago, when the Lion Guardians programme was still in its infant stages, only five Lion Guardians were employed to work around Amboseli, covering some 804 square kilometres of lion habitat. By 2017, Lion Guardians had trained and supported communities at seven different sites, successfully protecting lions around the continent. At their core site in the Amboseli-Tsavo ecosystem, close to 4,000 square kilometres of community land is now protected. There Guardians have reduced the killing of lions by local people by more than 95 per cent over the past decade, enabling the lion population to almost triple and be considered stable. In 2015, lions were spotted for the first time travelling between the Nairobi, Amboseli and Tsavo national parks.

Lion Guardians-based projects also operate in habitats across the border in and around the Ruaha National Park in central Tanzania and the Ngorongoro Conservation Area in the Tanzanian Crater Highlands. Communities in Rwanda and Zimbabwe have also used Lion Guardians tools and approaches to safeguard their lions too.

In Tanzania, records show that Lion Guardians-based programmes have reduced retaliatory killings by at least 90 per cent. For example, the Ruaha Carnivore Project based at the Ruaha National Park works closely with local Maasai and Barabaig communities and has employed warriors as Lion Defenders – trained by Lion Guardians – to protect lions. Just like the Guardians in Kenya, Lion Defenders earn an income and receive enormous respect from their local communities.

Since the Ruaha Carnivore Project was established in 2009 by Amy Dickman as part of a research fellowship with Oxford University's WildCRU, fifty lion hunts have been prevented or stopped. Lion Defenders have also reinforced more than 120 bomas, protecting around 16,000 heads of cattle – worth £1.7 million/US$2.2 million – from lions and other carnivores.

Year on year, in communities where Lion Guardians and the projects for which they have provided training operate, significantly fewer lions are being killed. This low lion mortality is the direct result of implementing and following practices that reduce conflict between lions and people. For example, the Lion Guardians' annual report for 2016 reported that the density of lions – four per 100 square kilometres of community land – had more than tripled since 2006. The report also stated that 189 bomas had been reinforced in the Amboseli-Tsavo ecosystem, and that 11,262 livestock (worth almost £1.5 million/US$2 million) that had wandered into the bush had been recovered by Lion Guardians. Twenty missing children were also found. And for the first time in living memory, cubs were seen in 2015 on the Ruaha unprotected rangelands, where the lion population is now no longer categorised as being in decline.

About the success of the Lion Guardians' work, Hazzah told the American news broadcaster CNN that she never imagined that warriors who were once lion killers would end up risking 'their own lives to stop other people from killing lions, but that is exactly what has happened'.

In Zimbabwe, down in the south of the continent, at the country's largest national park in Hwange, the Long Shields programme – a Lion Guardians-trained project – helps safeguard the 500 or so lions that live there. Named after Zimbabwe's famous Matabele warriors that once carried tall rawhide shields into battle, Long Shields are trained, just like East Africa's Lion Guardians, to help reduce conflict between lions and local people.

As in Kenya and Tanzania, the biggest threat lions face in Hwange is from local farmers who have lost cattle to lions that have roamed from the unfenced park into their villages. Shocked by the number of lions killed by local people, scientists and researchers attached to the Hwange Lion Research Project (run by WildCRU) set up the Long Shields programme in 2007 to help reduce the incidence of human–lion conflict.

Like the lions protected by Lion Guardians in Amboseli, many of Hwange's lions wear collars fitted with satellite tracking devices, enabling researchers to follow lions on the move. When lions are located outside of the park moving towards villages, Hwange's Long Shields are notified. Equipped with GPS trackers, mobile phones and mountain bikes, the Long Shields quickly head off to affected communities.

As well as state-of-the-art technological equipment, Long Shields also carry plastic vuvuzelas with them. Originally made from kudu horn and around half a metre in length, vuvuzelas are played like a trumpet, traditionally used to summon faraway villagers to community gatherings. Long Shields, however, have given them a new function.

When faced with unwelcome lions that have wandered into villages, the Long Shields blow their vuvuzelas. Often joined by villagers banging wooden blocks, sticks and drums, the racket they create is so awful that gate-crashing lions soon head back to the park, leaving livestock behind unmolested.

In October 2015, the *New Zimbabwe* newspaper described how Long Shield Charles Tshuma and a group of local people 'moved on' some wayward lions just 6 metres away from them by blowing their vuvuzelas and 'shouting and screaming until the lions turned away and ambled back to the Hwange National Park'.

Brent Stapelkamp, a researcher working with the Hwange Lion Research Project, also told the newspaper that this method of scaring away lions had proved so effective that conflict between lions and people around the park had reduced by 40 per cent since it was introduced. Vincent Mangenyo, a local leader, added that lion attacks had reduced and livestock were protected.

In northern Tanzania, where many of the country's most threatened lions live, the number of lions being killed in retaliation for livestock predations has been falling. 'Living Walls' have much to do with this. Living Walls are exactly what they say they are: they're walls made not from mud or bricks, but from living, breathing trees.

Traditionally, Maasai people have protected their livestock with bomas made from the thorny acacia trees that dot the dry plains of East Africa. Though initially hardy and strong, these woody barriers deteriorate and weaken quickly. Soon predators, not just lions but leopards, spotted hyenas and even honey badgers, break through the bomas and prey on the livestock inside.

Living Walls are different. They are strong, robust and resilient. Each Living Wall starts its life as chain-link mesh nailed onto posts, forming a circle in the ground. The posts are the slender branches of African myrrh trees, with their foliage trimmed off and just a few young, tender leaves left, dotted here and there.

The branches of myrrh trees are Lazarus-like. Once lopped off and planted in the wet season, they are washed into life, quickly growing roots and becoming reborn as new trees. Freshly grown branches interlink and wrap around each other tightly, creating an impenetrable barrier that no lion can break though. Like a safe, Living Walls are virtually impossible to crack. They protect from below too, with a root system so dense not even hyenas can tunnel through.

Living Walls are the brainchild of the conservation charity African People & Wildlife. Since 2006, when the charity was established, 650 Living Walls have been constructed, mainly in Tanzania but also in Kenya. Thanks to these walls, 125,000 livestock have been kept safe from lions and other predators. In places where Living Walls have been created, on average just one lion per year is now killed by herders, compared to six or seven before the myrrh trees were used as boma armour. Villages are now so keen to have Living Walls installed, there are waiting lists.

A similar initiative has also helped reduce conflict between people and lions. In the north of Namibia, the Kwando Carnivore Project has helped pastoralists make their bomas more resistant to lion assaults. Previously flimsy bomas made from acacia trees have been reinforced to keep lions out and panicked livestock in.

Since the bomas were strengthened in 2014, no livestock at Kwando have been taken by lions at night. Because of this, the number of lions killed by people has dropped from around twenty per year to just one in 2015. The news got even better in 2016, when no lions at all were killed by farmers in the region.

Although significant numbers of lions are killed by spearing or poisoning in retaliation for taking livestock, many are also killed by snares and traps set by bushmeat hunters and poachers. Bought cheaply or made quickly and easily from abandoned wire, snares are used mainly to trap herbivores, chiefly for selling as part of the burgeoning bushmeat trade.

Snaring affects lions in two ways. Firstly, it empties landscapes of antelopes, zebras and wildebeests, prompting lions to raid bomas for livestock to replace their vanished prey, leaving them to then face the spears, bullets and poisons of distraught herders. Secondly, many lions find themselves ensnared in the same deadly devices that garrotted their prey. Most lions die. A few survive, but are often maimed for the rest of their lives.

To help prevent lions from becoming snared, members of some conservation projects spend time in the bush looking for and removing poachers' snares. They also discourage local communities from making and setting them.

In the Luangwa Valley in eastern Zambia, where snaring is a significant cause of lion deaths, the Zambian Carnivore Programme and the South Luangwa Conservation Society have an anti-snaring team to help protect lions from traps. The team look for and retrieve set snares. Trapped lions are also rescued and given veterinary care and attention before being returned to the wild. In 2012, 357 snares were removed and 4 snared lions rescued, all of which were later released back into the wild after being successfully treated.

As well as removing snares, team members also focus on clearing up discarded wire in villages, which is used to make snares. Local people are also encouraged to store their surplus wire securely or dispose of it responsibly.

Many lion conservation projects are focused on research and play a significant role in reducing the number of lion deaths around Africa. By finding out more about how lions behave, researchers and scientists are able to better inform conservation initiatives that reduce conflict between people and animals.

Lion data, collected by scientists as well as by Lion Guardians and volunteers, reveals information about lion behaviour. It can include detail about how lions socialise and move around, the quality and size of their habitats, the abundance of food sources and their interactions with local people. Once analysed, this data helps guide long-term conservation strategies and policies that can help safeguard lions in both protected and unprotected areas.

At Zimbabwe's Hwange National Park, the Hwange Lion Research Project is involved in a major study led by Andrew Loveridge and David Macdonald from WildCRU. Using data collected from lion collars fitted with GPS, conservationists learned that, as expected, lions were targeting local cattle for food, but that this was something they only did during the wet season. During the rains, the park's herbivores stopped congregating around the few waterholes available in the dry season. Instead, they scattered far and wide, as they were now easily able to find water elsewhere.

As the herbivores disbanded, the park's lions found them harder to find and hunt. Rather than waste precious energy looking for scattered prey, Hwange's lions concentrated on another food source: prey much easier to locate in the form of livestock left unguarded by villagers while they focused on planting wet season crops.

Interviewed by Oxford University about his work, Loveridge said Hwange Lion Research Project's study showed that 'lions appear to weigh up the considerable risks of killing domestic stock by only doing so in times of wild prey scarcity'. Another of their projects, based at the Makgadikgadi ecosystem in neighbouring Botswana, showed lions doing the same thing.

As a result of these findings, local communities in both areas were advised to better protect their animals by strengthening their bomas and by providing more supervision of their grazing cattle during the wet season, when livestock are at their most vulnerable. Communities in Makgadikgadi and Hwange responded by building seasonal protective structures and guarding their animals. As a result, the number of livestock smash-and-grabs by lions during the wet season declined. At Hwange, for example, attacks on livestock by lions fell from 147 incidents in 2011 to 58 in 2014, a drop of nearly two-thirds in just 3 years.

Other research by the Hwange Lion Research Project showed that the park's lions were using wildlife corridors to move away from Hwange and into other protected areas. Once these corridors were identified, plans were put in place to ensure that lions could travel safely to them, enabling lions to establish new ranges and mate outside of their prides, encouraging long-term genetic diversity and healthier offspring.

Early in 2013, CNN and *The Guardian* ran a story about the invention of a thirteen-year-old Maasai boy who saved the lives of lions roaming around the Nairobi National Park in Kenya. Living with his family at Kitengela, just 32 kilometres south of the country's capital, Richard Turere had been looking after and protecting his family's cattle, sheep and goats since the age of nine. One of the biggest threats to his animals was lions breaking into their enclosures and killing them for food. Because of these attacks, Turere told CNN: 'I grew up hating lions very much. They used to come at night and feed on our cattle when we were sleeping.'

He also told *The Guardian* that his community believe they 'came from heaven with all [their] animals and all the land for herding them, and that's why [they] value them so much'. So when lions kill livestock, people take deadly revenge on the big cats, helping to explain 'why in the Nairobi national park lions are few'.

Around the Nairobi National Park, hundreds of lions have been lost to angry Maasai seeking revenge for the death of their livestock. Desperate to protect their remaining animals, villagers respond by poisoning not just the individual culprits, but sometimes entire prides too.

Aware of the decline in lion numbers, Turere was convinced that there must be a less violent way to keep lions at bay. First of all, he tried using a kerosene lamp and a scarecrow to keep lions away from his animals. But these efforts were unsuccessful, he told *The Guardian*, because 'lions are very clever' and soon realise that the scarecrow is not a person to be fearful of.

One night, though, while looking after his animals, Turere noticed that the flashlights of people moving around in the dark disturbed lions so much that they almost always ran away from them. Turere came to the conclusion that what he needed to keep lions away from his bomas was flashing lights.

He experimented by fitting flashing LED bulbs from broken flashlights to the poles of his boma, making sure the bulbs faced outwards and away from his cattle. He then wired up the lights to a solar-powered battery from an old car. His lights began to flash, and hey presto, Turere had found a method to keep lions away from his cows.

With the lights turned on every night, Turere's family had no more nocturnal lion visits and lost no more livestock. The big cats had been successfully duped into believing that someone carrying a flashlight was walking around the boma and stayed well away.

Seeing the success that Turere's lights had in keeping lions at bay, neighbouring families wanted lights for their bomas, too. As word spread, demand became so great that now more than 750 flashing-light rigs, known as Lion Lights, have been set up in Kenya and are saving lions' lives every night. 'I used to hate lions,' Turere told *The Guardian*, 'but now, because my invention is saving my father's cows and the lions, we are able to stay with the lions without any conflict.'

Lion Lights have provided an easy, innovative and low-cost way to reduce conflict between pastoralists and lions in a country that has lost thousands of lions killed in retaliation for attacks on livestock. Turere's invention has also helped support Kenya's wildlife tourism industry. Because there are more lions to see, there are more visitors to the Nairobi National Park, benefitting the local economy and communities.

The value of Turere's invention in helping to reduce the number of lion deaths was recognised by Paula Kahumbu, executive director of the Kenya Land Conservation Trust and chairman of the Friends of Nairobi National Park. She and her team were so impressed, they helped Turere get a scholarship to Brookhouse International School, one of the country's top educational institutes.

Some compensation schemes around Africa, notably in Kenya, have also played a role in helping to reduce conflict between pastoralists and lions. Compensation schemes, either state- or privately-run, provide farmers with financial payments to help offset the cost of buying new livestock when theirs have been taken by lions. In return for compensation, farmers and other community members must not kill any lions. Instead, they must tolerate the presence of nearby lions and learn to live with them. If any lions are then killed in revenge for livestock predation, compensation is withdrawn.

One of the most successful compensation schemes was started at the Amboseli-Tsavo ecosystem. It was set up in 2003 by the Big Life Foundation, a conservation organisation working with local communities in response to the dramatic fall in the region's lion population.

In an area of around 1,700 square kilometres, lions in the Amboseli-Tsavo ecosystem were, with fewer than ten remaining, almost extinct. The land these lions once roamed is traditional Maasai ground, where the vast majority of family incomes and livelihoods are made by raising livestock. As the Maasai population grew, lions gradually lost territories and prey, taking Maasai cattle and goats instead. In response, the Maasai killed so many lions that, over time, the number of lions left alive could be counted on the fingers of two hands.

To help protect Amboseli's remaining lions, the Big Life Foundation launched its first compensation programme, known as the Predator Compensation Fund, at the Mbirikani Group Ranch, home to a community of around 10,000 Maasai. The scheme offered partial compensation to Mbirikani herders who had lost animals to lions, as well as to other predators, including cheetahs, leopards, jackals and hyenas.

The impact of the scheme was positive and immediate. According to the Big Life Foundation website, once the compensation programme was in place, the number of predatory carnivores (including lions) being killed by the Maasai dropped dramatically – by 90 per cent. In the first six years of the project, just six lions were killed in retaliation for livestock deaths. This was a huge decline compared with the 200 lions killed during the same time period in neighbouring Maasai communities where no compensation was available.

As a result of the conservation efforts of the Big Life Foundation – as well as Lion Guardians and other NGOs, the Kenya Wildlife Service and most importantly the local communities who have left land and prey for lions – lion numbers at Amboseli have recovered. It's now one of the rare places in Africa where, on unprotected lands, lion numbers are actually growing rather than falling. Because conservation – in its many forms – is working, people and lions have found a way to coexist peacefully. These initiatives have been described as unprecedented successes in protecting lions and have been extended to support other Maasai pastoralists living nearby.

The Big Life Foundation compensation scheme has worked because it has been well managed and has received committed support from the local Maasai people it works with. To benefit from predator compensation funding, every single member of the community has to support the programme. If just one member of the community opts out and kills a lion, then compensation stops or is suspended for everyone.

The scheme's simple rules are made clear to the community and must be followed without exception. For example, when cattle have been killed by lions, their deaths must be reported within twenty-four hours and those animals must have been killed within around 1.5 kilometres of their home range.

Evidence of the kill must also be presented. All bomas must be reinforced and made as lion-proof as possible. Livestock are not allowed to wander around unprotected at night. If any lions are killed illegally by pastoralists and the offender is identified, a fine is issued.

And it's not just the lion killer that is adversely affected. All compensation payments are suspended for two months. So even innocent herders are impacted by the death of a lion. This form of collective punishment works well in Maasai culture, where causing problems and hardships for neighbours is particularly frowned upon.

Although compensation schemes can and do work, especially in times of drought when the number of livestock being killed by lions can escalate, they face challenges. One of the most pressing issues faced by compensation projects is funding. Few compensation projects in Africa are government-funded. Beleaguered by poverty and political uncertainty, few countries are in a position to earmark already limited and stretched resources for conservation and are unable to provide either financial or administrative support for such schemes.

This leaves most compensation schemes in private hands. They are often run by partnerships created by lion conservation charities and NGOs that are financially dependent on donations from individuals, groups and governments, generally based in Europe and North America.

Affected by global economic trends outside their control, forecasting long-term donations for compensation programmes is difficult and unpredictable. All that's certain is that when less funds are collected, the money available to provide compensation for affected herders becomes limited, sometimes even drying up altogether. This challenge currently affects the Big Life Foundation. In 2016, it published the following stark statement on its website: 'The funding for the Predator Compensation Fund is very shortly to run out of cash, leaving the fate of all the predators in the Amboseli ecosystem imperilled.'

The concern is that once compensation is withdrawn, lions taking livestock will be swiftly killed in revenge, just as before. Evidence has also suggested that lions are likely to be killed by herders whose compensation rates have been reduced or who were facing fines for breaking rules, and in times of extreme hardship as a way to gain the attention of the government. For some conservationists, the inability of compensation schemes to change long-term attitudes to living with lions is one of their biggest flaws.

Other issues that can affect the efficacy of compensation schemes are described by conservationists as the 'moral hazard' and 'perverse incentive'. The moral hazard describes the scenario where livestock owners deliberately provoke lion attacks to access compensation by adopting poor husbandry standards; for example, not strengthening bomas adequately against big-cat attacks.

To minimise and mitigate against this kind of behaviour, compensation projects, including those of the Big Life Foundation, insist that all bomas be fortified before any compensation is considered. However, as British conservationist John Murphy points out on the website of the lion protection charity LionAid, livestock owners 'usually hold the upper hand'. This is because 'they can always invoke the ultimate sanction of killing wildlife', while compensators must 'always be able to ensure that any penalty is enough to guarantee compliancy with agreements put in place but not harsh enough to provoke the ultimate act of retaliation by livestock owners who, at the end of the day, can simply refuse to cooperate'.

Managers of compensation schemes sometimes deal with the moral hazard by issuing compensation that is below the market price of livestock. However, working out what that rate should actually be is not simple. It needs to be high enough to discourage herders from killing lions, but not so high that farmers are tempted to lower their husbandry standards in order to access the funds. Deciding on the appropriate rate takes time, and when lions are vanishing quickly, there's often not the time to properly work this out and ensure the safety of lions by issuing the right rates.

The perverse incentive refers to the negative impact that compensation schemes can have on lions by encouraging communities to continue grazing livestock rather than consider other livelihood options. When people continue farming, lion habitat and prey continue to be lost, reinforcing the cycle of lions taking livestock and pastoralists killing them in response.

However, in places where compensation schemes are not offered, pastoralists may decide that the risk of losing livestock to lions is too great and may look for alternative work instead. For example, just west of the Masai Mara, along the Tanzanian border, hundreds of plots of land owned by individual Maasai farmers are no longer kept for cattle grazing.

Instead, they are rented out to tourism operators and safari camps who let the land return to bush, becoming new homes for lions and other wildlife, which tourists pay to watch. Local Maasai now earn income renting out their land, rather than grazing livestock on it; lions are no longer viewed as a threat to livelihoods and the retaliatory killing of lions has all but stopped.

As positive as this scheme is, it has its limitations. For the scheme to work, it depends on Maasai people owning individual pieces of land, something relatively unusual outside of the Masai Mara, where most Maasai land belongs collectively to large communities. But while compensation programmes like the Big Life Foundation's Predator Compensation Fund have their critics, they have saved lions' lives. At Amboseli, the killing of lions by local Maasai has almost stopped, leading Richard Bonham, co-founder of the Predator Compensation Fund, to tell *Time* magazine in 2010 that, yes, there had been setbacks, but that without the compensation project it was highly unlikely that any lions would have survived – it was compensation that put 'the brakes on the killing'.

Because compensation schemes have drawbacks, other initiatives have been considered to help pastoralist communities live in peace alongside lions. Some have found inspiration from a conservation model successfully used in Sweden.

In the north of Sweden, Sami herders receive payments not for the loss of reindeers taken by endangered wolverines and lynxes, but for the number of these predators with which they share the landscape. The more wolverines and lynxes that are counted year on year, the greater the herders' financial reward. Rather than being paid for dead reindeers, Sami herders are paid for the presence of living wolverines and lynxes instead.

In his 2015 book *Lions in the Balance*, lion expert and ecologist Craig Packer states his enthusiasm for performance payment projects that encourage pastoralists in Africa to better protect their livestock, practise high standards of animal husbandry and tolerate the predators that threaten their animals. For Packer, it makes more sense to pay people 'for the number of rare animals in their midst – rather than compensating them for lost livestock'. He sums up his viewpoint pithily: 'Hey, if you had a sick cow, wouldn't you just leave it out for the wolves to kill so that someone would buy you a healthy new one?'

Packer has spent a considerable number of years working to better understand and protect lions in Tanzania's Ngorongoro Conservation Area, up in the country's Crater Highlands. Home to thousands of semi-nomadic Maasai, much of this area is grazed by livestock, and retaliatory killings as well as ritual killings have adversely affected the local lion population.

Rather than introduce a compensation scheme to help reduce the number of lions killed by angry Maasai, Packer and his team decided to set up a performance payment scheme as a more effective and sustainable option, especially when also supported by a Lion Guardians-based programme.

After securing agreement from Maasai elders and the relevant wildlife authorities, Packer began to run a pilot scheme. To assess the number of lions living within the community's grazing land, camera traps were set up in local villages. Payments were determined by the number of lions captured by the cameras. Maasai living with more lions receive more funds than those who live with fewer lions. As lion numbers increase, so do the payments received by the Maasai.

Supported and trained by Lion Guardians, local Maasai men – known as Ilchokuti – help biologists collect camera trap evidence. They also discourage ritual lion killing, strengthen livestock bomas against attacks by lions and other carnivores, find and retrieve lost livestock and track lions, letting farmers know the areas to avoid grazing their animals on, essentially reducing the opportunities for conflict.

Packer is not alone in appreciating the benefits that performance payments can bring to both lions and pastoralists. Writing about performance payments in *SWARA*, the magazine of the East African Wild Life Society, Lawrence Frank, an experienced biologist who runs lion protection projects in Kenya, noted that if people were able to get more money as a result of having more lions, that would be considered 'a valuable asset rather than an expensive nuisance'. If this happened, land might be less grazed and wildlife might thrive. And to 'avoid funds disappearing as they pass through leaders' hands', monthly payments could be made via mobile phone banking on market day instead. This way, 'unsustainable dependence on foreign donors' could be reduced, since 'funds for such a programme' would 'come from the tourism industry and the government'.

Packer's performance payment project is still in its infancy, but it and other similar projects demonstrate the commitment there is in Africa for finding more and better ways to help people and the continent's last lions live together, free from conflict.

But what about the places where lions no longer remain, the places where they have been so persecuted that they have vanished altogether and are considered locally extinct? Although long gone in many instances, where conditions suit, lions have been successfully reintroduced, establishing new populations and safeguarding against the extinction of the species from the wild in general.

One place where lions have been reintroduced is the Akagera National Park in north-east Rwanda. Before their return in 2015, lions were last seen in the years following the genocide that ripped the country apart in 1994. As the bloody conflict subsided, thousands of returning, and now homeless, Rwandans who had been sheltering in neighbouring countries found land to settle on in the park, which had been abandoned during the chaos of the genocide.

People brought their cattle with them too, which soon became easy prey for the park's lions. In response, the lions were killed. In just a few years, all of Akagera's lions were gone. When the last lion roared at Akagera, Rwanda officially lost all of its lions. The country remained empty of lions for just over a decade until June 2015, when seven lions arrived at the park from KwaZulu-Natal in south-east South Africa. In 2017, I visited Rwanda to find out more about the reintroduction and its progress. And, of course, I also hoped to catch a glimpse of the nation's newest big cats.

Akagera National Park, Rwanda

I set off after breakfast with my 21-year-old guide Maddy Uwase. She is going to take me to the Magashi Peninsula in the north of the park, a highland region where the hills roll and lakes shimmer in the sun.

From the low-lying areas of the park's southern reaches, it's a journey of around two and a half hours. At the peninsula, there's a remote rangers' post where we will meet Nathan Mwesige, a ranger experienced in using radio telemetry and tracking the lions.

As we drive off, a cloud of red dust forms behind us. Acacia trees shiver slightly in the breeze. As the vista turns watery, Maddy points out Lake Shakani, just one of the ten lakes in the park, which together make Akagera the largest protected wetland area in Central Africa.

The papyrus that grows around the lakes reaches almost 3 metres tall in places. In such swampy areas, many bird species thrive, including rare shoebill storks, which forage for snacks in the form of baby crocodiles that have lost their way in the swamplands. Shoebills are just one of the 480 or so bird species that can be found at Akagera. Heading further north into the park, there are eagles – fish, hawk and bateleur – as well as smaller, more colourful birds like the exquisite lilac-breasted roller and the dazzling greater blue-eared starling.

It's not just Akagera's skies that are home to natural treasures. There's a richness of land mammals here too – more than 12,000 of them – including elephants, hippopotamuses, giraffes, buffaloes and recently reintroduced rhinos, as well as thousands of antelopes, from petite and delicate impalas to muscle-bound topis.

As we get closer to the peninsula, rain starts to fall. A telltale sign that the short, wet season that starts next month is just around the corner. Bright strikes of lightning stoke the sky and thunder ricochets around the hills. By the time we arrive at the rangers' post the rain is torrential, filling potholes in the road tracks, making them soupy with mud.

Nathan has been waiting patiently for us. He's based permanently at the post, a small brick building with a transmitter pole at its rear, seemingly in the middle of nowhere. As we get out of the vehicle, the rain stops and we greet Nathan. Swallows slip through the air around us, sky-dancing around the lofty transmitter as if it's a maypole.

Equipped with his receiver and antennae, Nathan hops into the jeep. While we've been driving, Nathan has been monitoring the lions and tells us that two males have been located and recorded roaming nearby.

Nathan suggests we drive towards Mohana Plain, a low-lying area, to see if we can find them there. At Mohana, Nathan raises his antennae. There is a faint sound of crackling and buzzing, like the last sounds of a fading bee. 'Are they here?' I ask Nathan excitedly. He shakes his head.

So we get back in the jeep and move on in search of the fugitive pair.

The lions we are searching for are part of a group now nineteen strong. The original settlers of seven – five females and two males – were part of a carefully orchestrated conservation project set up in 2010 by the Rwandan government and the conservation organisation African Parks (who manage Akagera through a public–private partnership) to bring lions back to Rwanda.

Translocated by plane and container truck, the seven big cats travelled around 3,218 kilometres, the longest distance that any wild lions have ever journeyed in Africa before. Accompanied by a team of veterinary professionals experienced in reintroducing African wildlife, the lions were tranquillised and monitored throughout their 26-hour-long journey, making it as trauma-free as possible.

On the evening of 30 June 2015, the seven lions arrived at Akagera in a truck driven from Kigali Airport, where the lions had been flown in from Johannesburg. As they passed over dusty roads to the park, the lions were welcomed to Rwanda by thousands of people. Schoolchildren with hand-painted banners and adults in brightly coloured clothes sang and cheered as they celebrated the return of lions to their country.

From their transportation crates, the lions were released into a temporary enclosure. Complete with its own water supply, the boma was the lions' first home in Rwanda, and where they were quarantined and left to adjust to their new surroundings. Observed by veterinary staff, the lions remained healthy and ate well. In just under a month, the seven lions were released from their boma into the wilds of Akagera. Rich grasslands, rolling hills, swamps and lakes were all part of their new kingdom, some 1,200 square kilometres in size.

After their release, the lions did what lions do. They started hunting immediately, their first prey a waterbuck brought down by one of the females. And it wasn't long before cubs

appeared – eleven of them in 2016. A year later, two more
male lions – adding to the genetic diversity of the big cats –
joined the original group.[*]

To date, there's been no conflict between local people and
their new feline neighbours. Before the lions arrived, a
community sensitisation programme was run with the help
and support of Kenya's Lion Guardians. Designed to help
allay any fears people might have about living near big cats,
the programme also highlighted the economic benefits that
lions could bring to local villages, as well as the region as a
whole. Many villagers also found comfort in the fact that all
of Akagera is enclosed by a predator-proof fence, creating a
barrier between local people and their lions next door.

Back in the jeep, Nathan tells me he's been working as a
ranger at the park for around two years. Both Nathan and
Maddy are local and have benefitted, like many others, from
the employment opportunities that Akagera provides.
Poaching is at an all-time low, and because of its reintroduction
initiatives, the park is now able to offer tourists all of Africa's
Big Five animals – lions, leopards, rhinos, elephants and
buffaloes. As a result, more visitors, from Rwanda and all
over the world, visit the park year-on-year.

After fifteen minutes or so, Nathan signals that we should
stop. He gets out and once again raises his antennae to the
sky. Nothing. We get back inside the vehicle. 'There's a
chance,' says Nathan, 'that the lions may be a little further
away.' The engine starts up and we drive on to an area of
scrubby terrain. A couple of metres from the jeep, Nathan
lifts his antennae skywards. He moves it from left to right and
back again. He does not give up easily. 'I want to find you
our lions,' he says, as he scrutinises his receiver.

Then Nathan's face clouds over. He looks crestfallen as he
tells me the lions have moved towards the mountains. There
the vegetation is inaccessible and impossible to pass through

---

[*] Sadly, one of the females died from natural causes while hunting
prey.

in a vehicle. And so, our search for the lions ends. We drop Nathan back off at the rangers' post and I drive back to the south of the park before night falls.

Of course I'm disappointed not to have seen the lions, but I am still delighted to be here. And it's a relief to know that the lions are just elsewhere rather than nowhere. Here at Akagera, they are safe, well protected and well managed.

I think back to the conversation I had yesterday with Jes Gruner, the park's manager, who led the lion reintroduction project. He told me that things are looking good for the lions. With its diverse and pristine habitats, Jes describes Akagera as a lion's paradise, and estimates that lion numbers here – well supported by the park's abundant prey – could triple with no problems.

Driving past stripy herds of grazing zebras, waterbucks with their soft donkey faces, crowds of impalas and little families of warthogs skittering on the road in front of me, it's hard to imagine lions going hungry here.

The lion reintroduction has been a conservation milestone, not just for Akagera but for Rwanda and Central Africa too. Many eyes are focused on the progress of the park's lions. They are a case study in progress, which may spur other former lion range states to welcome their big cats back so that children, like those living in southern Africa, can experience the joy of talking about their nation's lions in the present tense.

In Botswana, Namibia, South Africa and Zimbabwe, recent survey results show that lion numbers are up in the region by 12 per cent. In South Africa, lion populations are considered so stable that the next IUCN Red List will categorise them as being of least concern regarding the likelihood of their becoming locally extinct.

In Chapter Two, we saw that this hasn't always been the case for South African lions. By the 1860s, European settlers had completely wiped out the Cape lion that lived in the far south of the country. Four decades later, most of the lions that had ranged South Africa in their thousands were gone too, their lives lost to hunters keen to bag impressive trophies, or to farmers determined to protect their livestock and turn lion habitat over to agriculture.

So, what changed for South Africa's lions? How did they go from being locally extinct in some areas and scarce in others, to now being considered the best-protected lions in Africa? Historically, much of South Africa's resurgence as a haven for wildlife was triggered by changes in attitudes to wilderness and land use. First in the United States of America and then elsewhere.

From the 1850s onwards, thinkers and reformers around the world began to lobby for wild spaces to remain pristine, untouched by human activity. They advocated passionately that wild places mattered, that they enriched humanity and had value in their own right. Precious wild land and the wildlife that lived on it, they argued, should be officially protected by state governments in the form of national parks that could be enjoyed by everyone, now and in the future.

The United States moved first, establishing the Yellowstone National Park in 1872. Seven years later, Australia created the world's second national park, followed by the creation of parks in Europe from 1909. Africa's first national park, the Virunga National Park (originally known as the Albert National Park), was established in 1925 by Albert I of Belgium in what is now the Democratic Republic of Congo.

A year later, in 1926, the government of South Africa passed the National Parks Act and merged the private Sabie and Shingwedzi game reserves to create the Kruger National Park, the country's first national park, some 19,312 square kilometres in size.

Before national parks were established in the country, some areas of land had already been set-aside as reserves to be protected from development. Africa's oldest reserve, the Hluhluwe–Umfolozi Game Reserve (now a national park), was established in 1895 at KwaZulu-Natal in eastern South

Africa some thirty years before the creation of Kruger. However, these reserves were privately owned and in the main existed to protect game for hunters, since wild stock had become depleted as a result of over-hunting.

After 1926, the number of both parks and especially private reserves continued to grow. Private reserves were established by hunters, by conservationists, by business people, and sometimes a mixture of all three. However, they all shared the common goal of turning the land they owned back to a vista from yesteryear, where wilderness was reinstated and wildlife was abundant.

Reserve owners had big jobs ahead of them. Often, their land had been degraded by farming or had been developed and then abandoned. Much of the indigenous plant life and the herbivores that fed on it were long gone.

To turn the clock back to a time before Europeans altered the landscape, farmsteads were removed and non-native plants replaced with the original flora that had once thrived there. A similar approach was taken with wildlife. Exotic species that had been introduced were removed and indigenous animals reintroduced, taking their place. Some of the reintroductions were epic in scale.

At the Madikwe Game Reserve, close to the Botswana border, the world's largest wildlife translocation programmes were carried out between 1991 and 1997. Known as Operation Phoenix, more than 8,000 animals from 28 species were moved to Madikwe, an area of almost 805 square kilometres of restored land, once degraded and made barren by cattle farms.

Land denuded by grazing livestock and empty of the wildlife shot down by farmers and hunters was now home again to lions, leopards, rhinos, hippopotamuses, antelopes and deer. By resurrecting the land and reintroducing wildlife, a wilderness was reborn, a patch of old Africa recreated.

In conservation terms, establishing lost species at Madikwe was important for creating a new healthy and sustainable ecosystem, but there were financial imperatives too. For

Madikwe and South Africa's other reserves and parks to attract visitors from all over the world, it was essential to have Africa's Big Five most iconic animals – lions, leopards, rhinos, elephants and buffaloes – roaming the land.

After 1994, when apartheid ended and South Africa became a popular tourist destination, especially for wildlife lovers, it became even more important for parks and reserves to offer the Big Five, enabling them to maximise income through tourism. The economic value of animals, especially lions, was recognised early on and is still hugely valued today.

For example, at the Pilanesberg National Park in north-west South Africa, reintroducing lions as part of the Big Five was vital if it were to become a successful visitor attraction. Former park director M. L. Rammutla told the African Wildlife Foundation: 'Tourism companies believe that international travel agents will not promote or sell tours to African game reserves that do not have the Big Five.' It introduced twenty lions from Etosha National Park in Namibia in 1994. Now there are more than fifty.

In their 2014 article 'Translocations in South Africa: Lion Reintroductions in Perspective' in the journal *Current Conservation*, conservationists Matt Hayward and Michael Somers report that lions in South Africa are now safe from extinction quite simply because 'wildlife is privately owned and individual entities can make money from owning them'.

In effect, lions and other wildlife in South Africa are financial assets, and because they have monetary value they are kept safe – very safe. To avoid the conflict between people and lions that has caused lion numbers to plummet elsewhere in Africa, all private reserves and most parks are completely enclosed by fencing.

When lions are reintroduced, South African law requires that lion-proof fencing be put up. Any lions that escape must be immediately recaptured or destroyed, minimising the chance of local people and their livestock being harmed. Many conservationists believe that it's the fencing in of wildlife that has saved lions' lives in South Africa, preventing them from being killed as a result of conflict with people.

Pro-fencing conservationists argue that even though there are more lions living in South Africa than a century ago, it's exceptional to hear complaints of lions taking livestock – let alone human life. They also maintain that fences not only keep lions and other wildlife safe from people, they also offer a more cost-effective way to protect animals, since it's cheaper to manage animals within fences than those that roam outside.

For example, writing in the conservation journal *Ecology Letters*, Craig Packer and other scientists note that looking after lions in unfenced areas costs around £1,542/US$2,000 per square kilometre, compared to £385/US$500 for animals living in enclosed spaces. 'It is clear that fences work and unfenced populations are extremely expensive to maintain,' said Packer in the journal.

Although fences have kept lions in South Africa safe, some conservationists are against the fencing in of lions and other wildlife. While they accept that the practice has prevented significant amounts of conflict between people and lions in the south of Africa, they believe that the strategy jars with the belief held by many conservationists that protected areas should remain unfenced.

Critics argue that there are significant issues created by fencing, including habitat fragmentation, the disruption of migration routes and the narrowing of gene pools since lions can't intermingle beyond the fences to breed. Wire fences, they also point out, are often taken by bushmeat hunters and turned into snares, killing lions and other animals too.

As well as the protection provided by fencing, lions in South Africa have also benefitted from protection by well-trained wildlife rangers that patrol the country's parks and reserves. Ecologists working with wildlife and conservation managers also monitor habitats, keeping the landscape as natural as possible. Levels of prey are regularly checked to ensure that the balance of carnivores and herbivores is sustainable. Above all, the long-term welfare of lions is a priority.

In fact, in some smaller reserves, lions have done so well that there have been too many to be supported by existing prey stocks. Where this has happened, some lions have been given contraceptives to keep numbers in check. Others have been translocated and reintroduced to reserves that have no lions, or where fresh blood is needed to widen a restricted gene pool.

Recreating wilderness areas, then translocating lions and other wildlife, including elephants and rhinos, to fill them are specialist, long-term projects. Establishing a reserve takes time to plan and execute successfully, requiring the expertise and dedication of a whole host of professionals, including managers, conservationists, ecologists, biologists, veterinarians and other scientists. And it is, of course, expensive. Very expensive.

However, in South Africa both private and government investors have financially supported the establishment of reserves. Both sets of investors have appreciated the value of the country's natural heritage and how national parks and reserves form an important part of the nation's economic strategy. They are well aware that wildlife tourism brings far better returns on the land than pastoralism and farming ever could.

Because of this investment, around 6 per cent of South Africa's land is protected by the state, and a further 13 per cent protected by the private sector as game and eco-tourist reserves. Such investment, Hayward and Somers note in their *Current Conservation* feature, shows how South Africa 'illustrates to the rest of the world the value of adequately costing wildlife into national economics and performing active and intensive conservation management'.

South Africa has twenty national parks, the second-highest number in Africa after Kenya, which has twenty-three. All are looked after expertly and managed professionally by the government-supported South African National Parks

(SANParks), run through the Department of Environmental Affairs and Tourism. SANParks maintains high research and management standards and has expanded the land it protects at a rate considered unprecedented. According to the SANParks website, it now generates: '75% of its operating revenue – a spectacular financial achievement compared to most conservation agencies in the world, including those in developed countries'.

Of course, South Africa is one of the richest countries in Africa and able to invest in conservation in its national parks and reserves. It can afford to restock empty areas. It has the expertise to best protect its lions and other wildlife. And it has the money for fencing that stops hundreds of lions being killed by people.

South Africa is not only able but willing to invest in its natural heritage, in part because of its tourism legacy. As Africa's second most visited country, South Africa sees and feels the financial return on its investment. Because of this, the country's lions are the safest and best-protected lions in the continent.

While conservation projects all over Africa are finding practical ways to help save the continent's remaining lions, other individuals and organisations based inside and outside of Africa are lobbying governments and international bodies to achieve the same end.

Some of them believe that by banning the export and import of all parts from both wild and farmed lions, the number of lions killed, legally or illegally, can be significantly reduced. The world trade in wildlife is regulated by CITES (the Convention on International Trade in Endangered Species of Wild Fauna and Flora), an international agreement signed by more than 180 national governments, designed to protect more than 35,000 species of wild animals and plants threatened by global trade in their parts.

These species are protected under three appendices, each one demanding different degrees of export and import controls. Appendix I protects the world's most threatened species, including tigers and scarlet macaws. Trade in Appendix I species is banned and only permitted in rare and exceptional circumstances.

Species like the American black bear and the emerald-chinned hummingbird, which are vulnerable but not threatened directly with extinction, are listed under Appendix II. Trade in animals and plants listed in this appendix is not prohibited, but is strictly controlled.

Appendix III lists species, including the African civet and the golden jackal, that are protected in at least one country and depend on other CITES countries to regulate their import to support their survival. Species listed in this category can only be traded when relevant permits or certificates have been issued and secured.

Currently, lions in India are listed under Appendix I and African lions under Appendix II. For Asiatic lions, everyone agrees this is perfectly appropriate. However, many conservationists believe that African lions need the same degree of protection as lions in India and should be upgraded to Appendix I, making the global trade in all African lion trophies and parts illegal.

Moving flora and fauna to different appendices can only be done with detailed recommendations from CITES' member countries as well as the IUCN at CITES conferences. In Europe, conservationists lobbied the European Union to move African lions from Appendix II to Appendix I at the 2013 CITES conference in Bangkok, Thailand. The European Union declined to endorse the move, although it did recommend that the number of lion trophy hunting permits be reduced.

While plans to better protect lions in Africa faltered in Bangkok, it was hoped that *Panthera leo* would fare better at CITES' next key conference, held in Johannesburg, South Africa, in late September 2016.

There, a coalition of African countries, including Chad, Côte d'Ivoire, Gabon, Guinea, Mali, Mauritania, Nigeria and Togo, led by Niger, proposed that all African lions, wild and captive-bred, be moved to Appendix I and afforded full protection, effecting a complete trade ban in lion parts.

By upgrading the status of African lions, many experts maintained that the emerging trade in African lion bones, which has more or less tripled in the last decade, could be significantly reduced. For African lions to be transferred to Appendix I, recommendations for the move needed to be supported by two-thirds of the 182 countries at the Johannesburg conference. This didn't happen. Instead, a compromise was reached.

Panthera leo would remain listed as an Appendix II species, allowing hunters to continue taking home lion trophies from legal hunts. However, the commercial trade in bones, teeth and claws (but not skins) from wild lions in Africa would be banned. The trade in bones from captive-bred lions remained legal, although South Africa, the world's biggest supplier of lion bones, would now be required to submit an annual quota for their bone exports.

The news did not go down well with many conservationists, who feared that allowing a continuing legal trade in non-wild African lion bones would (as the WWF's wildlife trade specialist Colman O'Criodain pointed out to the BBC) 'keep the demand for big cat bone[s] alive, and complicat[e] enforcement efforts'.

Concerns were voiced that keeping the lion bone trade legal gave traders the opportunity to pass off bones from wild lions as those taken from captive-bred animals, and that other countries, in addition to South Africa, might look to set up lion breeding farms specifically to supply bones for export. Mark Jones, associate director of the Born Free Foundation, summed up the situation on the organisation's website: 'By continuing to allow trade in bones from captive-bred lions the Parties may have opened the door for products from hunted or poached wild lions and other endangered big cats to be laundered into trade.'

While many delegates in Johannesburg were disappointed by the rejection of the proposal to upgrade Africa's lions, there were member countries that were pleased with the ruling, maintaining the view that funds raised from regulated legal lion hunting could support conservation initiatives dedicated to protecting the species.

Zimbabwe's representative, for example, noted that in many African countries, including his own, lions were killed in retaliation for livestock kills and that the practice would only end if value was associated with lions 'through eco-tourism and sport hunting, with the money ploughed back into conservation'.

Although CITES plays a hugely important global role in regulating trade in wildlife around the world, individual countries can themselves dictate what plant and animal species are to be traded on their shores. Outside of CITES, Australia, France and the United States have all placed legal restrictions on the import and export of lion trophies brought back from Africa by their nationals.

In March 2015, Australia, which had from 2010 to 2013 imported the parts or bodies of ninety-one lions, became the first country in the world to ban the import of all lion trophies. By making it illegal for hunters to bring home their spoils, Australia hoped to protect lions and other African wildlife from organised hunting. Hunters found guilty of bringing lion trophies from Africa into Australia face ten years in prison, as well as fines totalling hundreds of thousands of Australian dollars. The Australian ban was prompted by an anti-trophy hunting campaign that attracted huge public support after photographs showing the country's famous test cricketer Glenn McGrath posing with African wildlife he had killed and shot in Zimbabwe went viral.

Eight months later, in November 2015, France followed Australia's lead, becoming the only country in the European Union to ban the import of lion parts, including skins, paws and heads. On the 19 November, Ségolène Royal, France's environment minister, declared publicly that the country, which had allowed the import of more than a hundred lion trophies from 2010 to 2013, would no longer be issuing permits allowing lion parts to enter the nation.

A month on, in December 2015, the world's biggest importer of lion trophies, the United States, officially added African lions to its Endangered Species Act, making it much more difficult for hunters to bring lion trophies back to the States. Although Americans can still legally hunt lions in Africa, bringing home a head, skin or other part is no simple matter. To import a lion trophy under the new ruling, hunters must source a permit from the country's Fish and Wildlife Service (USFWS).

However, securing a permit relies on meeting specific conditions. USFWS provides two different permits: one for endangered lions living in India and Central and West Africa, and another for lions from the east and south of the continent, listed as threatened. To obtain a permit to bring back parts from endangered lions, hunters must provide documentation from their host countries proving that their lion conservation work is scientifically sound. Also, to bring back trophies from threatened lions, hunters must prove that their hunting fees have been used to help the survival of the species by supporting lion conservation projects.

Regarding the move, Hans Bauer, a WildCRU lion conservation researcher, told the *New York Times* that the burden of proof had now shifted, since under the 'new ruling, countries must not only prove that hunting is not bad for lions; they must prove that it is good for lions'.

Providing such proof isn't easy. Project evidence must be robust and transparent. It must show high levels of scientific management and it must be effective, proving that lions really are benefitting. As well as helping to ensure that conservation projects funded by hunting fees are legitimate, it's also thought that the more complicated application process might put some hunters off altogether.

Some ten months later, USFWS made a statement that shocked many hunters hoping to hunt lions in South Africa. On 20 October 2016, after the CITES conference in

Johannesburg, USFWS director Dan Ashe announced that new legislation now made it illegal for Americans to bring back trophies from captive lions killed in South Africa. Writing in the *Huffington Post*, Ashe stated that for lion parts to be legally imported into the States, South Africa had to demonstrate clearly the 'conservation benefit to the long-term survival of the species in the wild', and that 'in the case of lions taken from captive populations in South Africa, that burden of proof has not been met'.

He also added that the decision in favour of the ban was based 'solely ... on [their] evaluation of the conservation benefits of captive lion hunts'. Since there was no evidence supporting the long-term survival of lions in the South African wild, no permits would be issued to American hunters hoping to shoot captive-sourced lions.

In a similar vein, Ashe added that USFWS had received applications from hunters based in the United States planning to hunt in Mozambique, Namibia, Zambia and Zimbabwe, and to secure 'permits to import sport-hunted lion trophies'. Ashe noted that, in response, USFWS was duly assessing hunting programmes in these countries and would only be approving applications where the long-term benefits to lion populations were indisputable.

In September 2017, the journal *Nature Conservation* published the review article 'International Law and Lions (*Panthera leo*): Understanding and Improving the Contribution of Wildlife Treaties to the Conservation and Sustainable Use of an Iconic Carnivore', written by international wildlife lawyers Arie Trouwborst and Melissa Lewis from Tilburg Law School in the Netherlands, lion experts David Macdonald and Amy Dickman, and other scientists from WildCRU.

The review assessed the role of international treaties and conventions, including CITES and World Heritage, in lion conservation. While noting the positive contribution that such bodies make to their safeguarding, the review also considered ways lions could be protected through further, more interconnected legislation. One important recommendation

Trouwborst, the review's lead author, and his team made was for lions to be formally listed under the Convention on the Conservation of Migratory Species of Wild Animals (CMS) – a global platform for the conservation of migratory animals and the lands they live on.

CMS works by bringing together the countries that migratory animals pass through (known as range states) and by providing the legal framework for internationally co-ordinated conservation measures within migratory ranges. According to Trouwborst, adding lions to CMS Appendix II, which calls for international co-operation to ensure that the conservation status of listed species is positive, 'made perfect sense because not only would it raise the profile of lions', it would 'moreover enable the CMS to provide a framework for co-ordinating and assisting conservation efforts in the 25 countries where lions still occur in the wild'. A month or so later, at the 2017 CMS summit held in the Philippines, the proposal was given the green light and lions were officially added to Appendix II.

Almost four weeks after Cecil's death on 28 July 2015, popular US chat show host Jimmy Kimmel broadcast an impassioned and heartfelt speech lamenting the lion's demise and lambasting trophy hunting. Kimmel spoke of the brutal way in which Cecil had been killed and of the shocking decline in lion populations in Africa in general. He urged people to help save African lions by donating to Oxford University's WildCRU research project in Zimbabwe, of which Cecil had been a part of.

Listening intently, Kimmel's shocked viewers sat up and took notice. More than 4.4 million of them went online that same night, crashing both the WildCRU and Oxford University websites, and donated around £500,000/ US$647,412. At the same time, Cecil's death went viral too, igniting an unprecedented global media reaction to a wildlife story in the span of just two days.

In articles and columns in newspapers and magazines around the world, mentions of Cecil peaked at 11,788 on 29 July. On YouTube, Facebook and Twitter, Cecil was even more popular, peaking at a staggering 87,533 mentions. Celebrities, including the comedian Ricky Gervais, tennis star Andy Murray and supermodel Cara Delevingne, tweeted their outrage about Cecil's death and made their abhorrence of trophy hunting known.

All over the world, including Australia, Africa, the UK, North America and parts of South America, thousands tweeted their own thoughts about Cecil. An online petition, 'Justice for Cecil', was signed by close to 1.2 million people, lobbying the Zimbabwean government to ban the trophy hunting of endangered animals.

Such was the global coverage that David Macdonald described the public outpouring of feeling for Cecil as a 'unique moment in conservation' and arguably 'the biggest global response to a wildlife story there's ever been'.

In addition to signing online petitions, funds were raised, vigils were held and marches were attended to mark Cecil's death and raise awareness about the desperate status of Africa's lions. Responding to the public mood, forty-one airlines, including Air Canada, American Airlines, British Airways, Emirates, KLM, Lufthansa and Qantas, reaffirmed their commitment or newly committed to stop carrying lion trophies, as well as trophies from the other Big Five species.

Taken aback by the intense interest in Cecil, WildCRU recruited a media monitoring company to help them understand the wider impact that Cecil's death might have for lions and conservation in general. In his WildCRU paper considering the findings of the media monitoring company, Macdonald considered:

whether the preoccupying interest in Cecil displayed by the millions of people who followed the story may betray a personal, and thus potentially political, value not just for Cecil, and not just for lions, but for wildlife, conservation and the environment. If so, then for those concerned with how wildlife is to live alongside the human enterprise, this is a moment not to be squandered and one which might have the potential to herald a significant shift in society's interaction with nature.

Could it be, that alongside the work of on-the-ground conservation projects, lobbyists and legislators, a groundswell of public opinion could also be harnessed to help ensure *Panthera leo*'s future?

CHAPTER SEVEN

Beyond Cecil

The Summit holds the potential to mark an extraordinary
turning point for lion conservation. As the participants
showed, the land and people are there to save lions. What is
needed is will ... and commitment.

Thomas Kaplan, founder of Panthera

The Cecil Summit, Oxford University, Oxford, England
It's one of those balmy golden evenings that September
sometimes treats us to. I'm making my way to a lecture hall at
Oxford University's Blavatnik School of Government for the
final session of a conservation summit dedicated to saving
Africa's last lions.

Organised and hosted by the university's Wildlife
Conservation Research Unit (WildCRU) and Panthera, the
wild cat conservation organisation, the Cecil Summit is being
held in memory of the lion known as Cecil that was shot and
killed in Zimbabwe by American trophy hunter Walter
Palmer in the summer of 2015.

The world's leading lion experts have been invited here
with the objective of creating a twenty-first-century
conservation plan that will help end the catastrophic collapse
of wild lion populations in Africa.

I arrive at the lecture hall a little early and quickly find a
seat. As photographs of Cecil are projected onto the wall at
the front, the hall starts to fill with men and women who
have travelled from all over the world to be here. Some have
come from Africa. Others from Scandinavia, Europe and the
United States, as well as Scotland and elsewhere in the UK.

Some are university students, fresh-faced and still learning about conservation. Others are more experienced and include seasoned conservationists with hands-on involvement helping African communities and lions live together. There are dedicated scientists working on research projects, gathering data about lion behaviour. Representatives from the BBC's natural history unit and the specialist cat group at the International Union for Conservation of Nature (IUCN) are here, too. But whatever their background, everyone shares something in common. Lions. They all desperately want to save Africa's last wild lions, and are keen to work together to devise a new way to best achieve this.

As it gets closer to 5 p.m., when the session is due to start, the sound of excited chatter hushes to near silence. Social media apps are closed and mobile phones turned off. From the back of the hall, WildCRU director David Macdonald walks down to the front, welcomes us and starts his presentation.

He begins by summarising the plight of African lions, noting their dramatic decline from more than 200,000 a century ago to around just 20,000 today. Next, he speaks of WildCRU's research project in Zimbabwe and how Cecil, after roaming into their study area in Hwange National Park in 2008, became one of the lions studied by the project's scientists.

While Macdonald speaks, more photographs of Cecil, black-maned and magnificent, are projected onto the wall behind him. The final picture is poignant, and probably the most well-known. It's the last image we have of Cecil. It shows the lion's huge head hanging lifelessly, his eyes closed tight, as if in pain. Behind Cecil's splayed-out body, Walter Palmer kneels and grins, bright-eyed and pleased with himself.

It's the gruesome image that appeared in thousands of tweets and other social media posts all over the world. Macdonald says it was part of the online phenomena that caused Cecil's death to go viral, crashing websites and generating thousands of pounds for lion conservation as it did so.

It was, Macdonald says, a key moment in wildlife history, referred to as 'the Cecil Moment', which showed that people care passionately about lions and their survival. Which is why the Cecil Summit has been set up – to ensure that the lion's death was not in vain and that the Cecil Moment will, with the ideas and commitment of the summit's delegates, become a movement that keeps the world's attention focused not just on the future of Africa's lions, but on all of Africa's wildlife.

The summit is a conservation turning point, where both lion experts and specialists working in other fields have come together to create a new strategy, an out-of-the-box blueprint, a breakthrough that will save Africa's wild lions from extinction.

The public lecture I'm attending is the summit's final event. Earlier sessions have taken place in Tubney, a small village around 14 kilometres from Oxford, at WildCRU's Recanati-Kaplan Centre. Close to sixty delegates from all over the globe have been involved in discussions, debates and workshops – brainstorming ideas in an attempt to break the traditional conservation mould and come up with new ways to end the current crisis in lion conservation.

Significant advances have been made in the day-to-day protection of lions in Africa, especially in reducing conflict between lions and people. Present projects are doing a fantastic job in themselves, but aren't able to protect lions beyond their territory. The focus of the Cecil Summit, says Macdonald, is to look beyond the projects, which he describes as 'marvellous sticking plasters'.

In Tubney over the past few days, renowned conservationists, biologists and scientists, as well as community representatives from the countries where lions live, have been putting their heads together with well-regarded professionals from outside the field, including 'influential experts from … international policy, law enforcement, communications, economics, ethics, law and other sectors', to find innovative ways to prevent further collapse of the wild African lion population.

Their remit has been to find 'extreme interdisciplinary' solutions 'across the socio-economic and political landscape', aiming to set what Oxford University and Panthera describes as a new 'precedent for conservation' to help secure the survival of *Panthera leo*, so that 'the voice that was given to Cecil and his species may yet turn into a roar'.

Macdonald tells us this is an approach so groundbreaking that, perhaps at future conservation summits, institutions will use the same method and refer to their event as having used the 'Oxford format'.

Next, Macdonald introduces the panel who have professional and personal backgrounds as diverse and high-calibre as the summit's delegates. Moderating the discussion between panel members and the audience is Alan Rusbridger, former editor of *The Guardian* and now principal of the university's Lady Margaret Hall.

The panel itself includes Lovemore Sibanda, community liaison officer and Long Shields project co-ordinator for WildCRU's lion research team at Hwange National Park in Zimbabwe, and John Kamanga, a Maasai community chief leader working for the Southern Rift Association of Land Owners, a Kenyan-based land trust.

They're joined by Brazilian-born Achim Steiner, head of the United Nations Environment Programme, and via Skype, Rory Stewart, former UK minister of the environment, now minister for international development. Finally, from the United States are Craig Packer, lion biologist, author and director of the Lion Research Center at the University of Minnesota, and Thomas Kaplan, Panthera founder and chairman.

The session is then formally opened with the screening of a video message from Edmond Moukala, Africa unit chief at UNESCO. In the message, Moukala expresses UNESCO's support for the summit and for saving Africa's wild lions – animals regarded by UNESCO as a heritage species, one whose survival is important for the enrichment of both the natural and cultural worlds.

After Moukala's address, Macdonald summarises the three things that need to happen to reverse the current decline in lion numbers. Firstly, he says, it's imperative to better protect the places where lions still live. Secondly, people – not just those living in lion range states but in nations all over the world – need to be inspired to get involved in saving lions. And thirdly, there's an urgent need to secure finance from wealthy countries to fund lion conservation in the economically challenged African countries where lions still roam.

During the three days of the summit (4–7 September 2016), Macdonald explains, delegates have been brainstorming ways to deliver these three calls to action. By adopting an interdisciplinary and holistic approach to lion conservation, delegates have responded by establishing five 'headline' recommendations to bring the king of the beasts back from the brink of extinction. He then points out that, although none of the recommendations are unique in themselves, their interdisciplinary and holistic approach is new and 'breaks the mould of lion conservation'.

Before sharing the delegates' five headline areas, Macdonald stresses that each solution will remain part of an ongoing conversation involving not just lion insiders and experts from other professions, but also people who share their landscapes with lions, whose input and contribution is equally vital for saving the species.

Macdonald then shares the five interlinked headline areas identified by the summit's delegates, as follows.*

1. Restore lionscapes
Although human encroachment is a significant threat to areas where lions live (lionscapes), Africa still has 1.2 million square kilometres of viable lion habitat. Lions currently occupying protected areas have plenty of room to roam – so much so

---

* Additional detail has been added to each area since the summit took place. This extra information was made available by WildCRU and is detailed at whenthelastlionroars.co.uk.

that there is capacity in these areas to accommodate significantly more lions than Africa is currently home to.

Of course, this is good news, but without action now to keep these areas intact, they will not survive the inevitable human encroachment that accompanies rapid population growth. Ensuring the long-term sanctity of protected lionscapes is paramount for preventing further lion loss.

Secure the six and save the sixty

Currently, there are 6 protected areas* distributed among 7 countries – Botswana, Mozambique, Namibia, South Africa, Tanzania, Zambia and Zimbabwe – that support populations of more than 1,000 lions. Known as strongholds, these areas are home to the largest groups of lions on earth.

By focusing initially on these six lionscapes – restoring land as well as providing greater protection and more effective management – lions here will have the best chance to grow in number. As lions start to thrive in these places, so too will other wildlife, including elephants and other threatened species.

Once lion populations in these six key areas grow larger, then the focus can move to the fifty-four areas where other lion populations still exist but in much smaller numbers – half of them with fewer than a hundred individuals – and whose current long-term survival is unlikely.

2. Inspire national communities in lion range states

For many African communities, living with lions is hugely challenging. Most don't receive any economic benefit from doing so. Instead, many find their lives ripped apart by grief and fear when family members are killed or maimed by lions. Others find their livelihoods devastated when lions kill their cattle and goats and they can't afford to buy new stock.

---

* Details of the six areas identified by WildCRU are available online at whenthelastlionroars.co.uk.

For affected people, especially those living in Africa's poorest countries, lions often have no intrinsic value. Rather, it is lions that devalue their lives and livelihoods. As a consequence, thousands of lions are lost by being speared and poisoned and their habitats appropriated. To reduce the number of lions killed this way, attitudes and behaviours towards lions need to change for the long term. Mechanisms need to be devised that instead inspire countries with lions to yearn to keep them, and for those without them to actively want them.

Social justice

One way this can be done is by creating fairness in conservation activities, so that the costs don't fall at the doors of the people least equipped to cover them. The economic advantages that lions can bring need to be maximised and shared so that local people receive financial benefit, too. Ensuring that lionscape communities are genuinely better off as a result of the presence of lions can change negative attitudes towards them.

Tailored 'payments to encourage coexistence' (PECs) need to be developed to encourage the people that live with carnivores like lions to 'want the long-term conservation of globally iconic but locally problematic species', says Amy Dickman, lead scientist of the paper 'A Review of Financial Instruments to Pay for Predator Conservation and Encourage Human–Carnivore Coexistence', published in *PNAS* in 2015.

PECs, she says, need to meet both the cultural and economic needs of the people that carry the real cost of coexisting with lions. And they need to be adjusted to suit everyone – from very poor cattle herders to prosperous farmers. The perfect PEC scheme must provide all local people with 'tangible, commensurate, and equitably distributed benefits from predators that outweigh all the diverse costs, and carnivore-related revenue can help people escape existing poverty traps'.

When people find that living with lions can boost their income rather than threaten their livelihoods, lions are no longer seen as vermin that only make money for tourism operators. Instead of killing lions, people prefer to protect them.

As well as ensuring that lions have a clear economic value for the people who live with them, lions also need to be valued as a globally iconic species with a rich cultural heritage – the stars of countless histories, myths and legends, and the subject of famous artworks around the world. But they also need to be valued for just being what they are.

An opinion piece published in August 2017 entitled 'Some Essentials on Coexisting with Carnivores', authored by John Vucetich (a visiting scientist at WildCRU) and David Macdonald, expands on this concept. It states that animals are valuable not only because they can support the well-being of people, but because they have intrinsic value in their own right: 'As such, nature's utility is an (important, but) insufficient motivation for conservation.'

For the conservation of large carnivores to be successful, people must not be solely motivated by personal gain but must also want to save them because they are worth saving in themselves, so that animals like lions 'are not just chips in the game, they are stakeholders in it'. When communities feel proud of the prides with which they share the land, they regard the survival of their wild lions as imperative and view their loss, both locally and to the world, as unacceptable.

3. Inspire a global community

If lions are to survive, it's not just countries with lionscapes that must value lions. Millions of people all over the world must desire the long-term survival of the species too. After Cecil was killed, there was a global sharing of grief and rage. Thousands of pounds were raised for lion conservation. However, not enough was raised to safeguard all of Africa's lions.

The interest raised in the plight of lions since Cecil's death must be mobilised, maintained and grown. Something that Macdonald is committed to, and delivers by maintaining contact with 14,000 people who donated to WildCRU through updates and video messages, helping to keep people interested and involved in lion conservation.

However, for wild lions to avoid extinction, further awareness of the perils they face need to be put in the world's spotlight: the Cecil Moment must become a lasting movement. A hard-hitting and powerful campaign needs to convert people's concern for the future of lions into regular, committed giving to help fund the restoration of Africa's six largest remaining lionscapes.

4. Enact the Robin Hood model

In the UK, the European Union and the United States, national parks are funded largely from taxes raised by national governments. In Africa, very few parks are protected this way. Their funding comes from entrance and trophy hunting fees. Typically, the income they generate is not enough to provide proper protection for lions and other wildlife, and rarely benefits local people financially.

As a result, animals are poached, habitat is lost to encroachment and local people are not protected from wandering carnivores. Conservation, especially lion conservation, is expensive. Anti-poaching reinforcement, infrastructure, management and human–wildlife conflict mitigation does not come cheap.

For example, a study published in the journal *Biological Conservation* in 2017 called 'The Performance of African Protected Areas for Lions and their Prey' showed that most protected areas in Africa are chronically underfunded. And that as a consequence, 'just 31 percent of protected areas with lions currently maintain the species at 50 percent or greater of the natural density they would reach if only suffering natural mortality'. For lion conservation to work, it needs to be adequately funded by the world's wealthiest countries, not by the poorest people living in the poorest nations on the planet.

The affluent global community must step up and provide additional, significant and long-term financial resources to

Africa's lion range states, enabling them to restore the continent's lionscapes and the lions that live in them.

5. Fair international financing of lion conservation

Currently, most lion conservation is financed, albeit inadequately, by African nations already struggling to survive economically. And yet, the whole world enjoys and appreciates the lions they are able to protect. It's a global benefit funded largely 'by the poorest countries in the world, and sometimes by the poorest people in those countries', say WildCRU.

For example, a study published in the journal *Global Ecology and Conservation* called 'Relative Efforts of Countries to Conserve World's Megafauna' created a Megafauna Conservation Index that showed how much 152 countries spent on protecting their large mammals. It showed that many developing African nations made biodiversity more of a priority than those living in more affluent countries.

According to the index, 90 per cent of countries in North and Central America and 70 per cent of countries in Africa were classified as above average in their efforts to protect their megafauna. Even though many countries in Africa face poverty and political instability, the continent still contributed 'more to conservation than any other region of the world'. Botswana, Namibia, Tanzania and Zimbabwe topped the list, while the well-off United States was ranked 'nineteenth out of the twenty top performing countries'.

To raise the finances required to adequately fund lion conservation, international governments in the UK, the European Union and the United States, as well as international agencies like UNESCO, the United Nations Environment Programme and the World Bank, must get involved. They must appreciate the scale of long-term funding needed to save a species and provide it accordingly.

At the close of June 2016, Panthera, WildCRU and WildAid (a charity dedicated to ending the illegal trade in wildlife) published a report called *Beyond Cecil: Africa's Lions in Crisis*. It considered the factors causing the free fall in lion numbers. To help bring back lions from the brink of extinction, it concluded that at least £1.5 billion/US$2 billion per year is needed. This money would enable the six key protected areas in Africa to be effectively managed and saved from poaching and human encroachment, enabling lions and other wildlife to thrive.

Prohibiting trophy hunting was not put forward at the Cecil Summit as a 'headline' area. Although the activity can adversely impact some lion populations, overall trophy hunting is not considered a major threat to lions across Africa as a whole. However, there is a strong consensus that trophy hunting must be urgently reformed throughout the continent, especially in areas where lions are already at risk from human activity, including poaching and conflict over livestock.

In their *Beyond Cecil* report, Panthera, WildCRU and WildAid propose unilateral reform that must be monitored and enforced, specifying penalties when regulations are broken. Some of the measures proposed to ensure lion populations are not put at risk by trophy hunting include the following:

- That harvests be restricted to 0.5 lions per 1,000 square kilometres, except in the 'very best managed areas'.
- That the minimum age a lion can be harvested be seven – rather than five, the current minimum age.
- That quota setting be scientifically conducted, site-specific and transparent.
- That central governments be involved so 'they can devolve user rights over wildlife to communities in the hunting blocks where people live, so that they can derive most or all of the benefits from hunting'.

Other necessary actions include the rooting out of corruption, ramping up anti-poaching activities, increasing community outreach work and introducing the certification of operators.

Similarly, canned hunting (when captive-bred lions are shot by paying hunters in an enclosed area and the hunters are guaranteed a trophy) was not addressed as a headline issue at the summit, as it is not considered a major driver of lion declines. However, WildCRU and Panthera have stated that there is a case for its prohibition, in so far as it already has – and may increasingly have – a detrimental effect on the conservation of wild lions. Advocates of canned hunting say that the activity removes the pressure on wild lion populations because hunters are only killing captive-bred animals. They also maintain that being able to source lion bones legally from a captive-bred source (South African canned hunting operators) protects wild lions from poachers.

However, opponents point out that the hunters who pay to hunt captive-bred lions are a different market to the hunters that want to take part in a 'fair chase'. Canned hunting, they say, does not reduce the number of wild lions killed by trophy hunters. They also stress that being able to source lion bones and parts from captive-bred sources does not help wild lions because many consumers believe that wild lion bones are more potent than captive-bred ones, and are not interested in purchasing the products that contain them. Inevitably, wild lion bones will join the same trade routes as captive-bred lion bones, say WildCRU.

After sharing these headline areas, Macdonald invites the panel and audience to respond with their thoughts and ideas. Two hours of engaged and passionate debate follow, skilfully steered by Alan Rusbridger. Both panel and audience members talk eloquently about the ethics of conservation, the lessons learned from working and living with lions, and the urgent need for hard-core action to finance lion conservation via a single dedicated global fund.

From the audience, Joe Mbaiwa, a tourism expert from the University of Botswana, points out that an arrangement directly involving African people is vital to helping lion numbers increase:

> We should promote interaction between all the stakeholders and come up with a bottom-up approach in lion management, instead of only scientifically focused solutions that come from experts in universities.

Similarly, John Kamanga notes:

> We have to marry conservation ideology with the practices of pastoralist communities.

Fellow panel member Achim Steiner stresses the importance of sharing the economic value of lions directly with the people who live with them:

> Why should a nation care about keeping lions … they take up a lot of land and they kill cattle and people … [But] for the farmer in Botswana for whom lions cost a lot of money, there is another person in the United States willing to pay up to US$2,000 a night to sit next to that lion. If you can connect these two realities, you begin to have a source of economic benefit.

In the audience, Kirk Hamilton from the London School of Economics, and formerly the World Bank, adds:

> In some African countries, 3 to 6 per cent of their GDP derives from tourists who come to see wildlife, so conservation in Africa is a developmental issue.

By acknowledging that conservation projects can have developmental impact and help people move out of poverty, the scope for funding the preservation of lion habitat in

Africa also becomes wider and more likely. Supporting this
way of thinking, Rory Stewart comments that his instinct in
'saving 1.2 million square kilometres of lion habitat in Africa'
could form part of an international development project as
Britain has the resources and 'it is an issue that's dear to
people's hearts'. That a British government minister and
indeed a minister for international development should make
such a statement has been regarded by some as historically
important.

The conviction that the crisis in lion conservation can be
turned around feels real. It feels strong. The session closes
with a heartfelt and optimistic rally to action from Thomas
Kaplan. He talks about Panthera's Jaguar Corridor Initiative
currently being implemented in fourteen of the eighteen
countries where the near-threatened big cat still roams. He
reveals that the initiative has inspired national governments
in Latin America to value their jaguars and work together to
save them.

Kaplan also refers enthusiastically to the work of WildCRU's
Hwange Lion Research Project, whose efforts convinced the
Zimbabwean government to reduce its lion-hunting permits
from sixty to just four a year. A move that saved Hwange's
lions from likely local extinction. 'That's conservation, baby!'
he concludes, to the delight of the audience. 'That's the gold
standard.' An atmosphere of hope and positivity fills the
lecture hall.

We've learned that Africa has enough protected land for
lions to more than double in number. Lions aren't crowded
out just yet. A new interdisciplinary roadmap created by
some of the finest minds around can turn the Cecil Moment
into the Cecil Movement, and generate enough funds to
protect lion habitat from the impact of forthcoming human
population growth. The African lion's story is not over. It
could still be good to be the king.

But what next? What are the practical things that need to
happen to make lion conservation successful in the six key
lionscapes? With the right amount of funding secured to

protect such prime lion habitat, it's feasible that lion numbers could double as soon as 2036. But, how can this funding be secured?

Some of these answers come after the summit. Speaking to *National Geographic* magazine in October 2016, Craig Packer outlined the plan to effect real change, not just for lions but for the other iconic African animals that also live in lionscapes.

The first priority, Packer says, is to work more closely with the relevant authorities and local communities in lion range states. Growing their commitment to saving their lions is vital. Top-level meetings with national leaders will be held to secure robust agreements to reinforce protection of lions and their habitat. Their links with conservation organisations outside Africa will also need to be strengthened 'to build a global platform for protecting their imperilled National Parks and Game Reserves'.

To fund their conservation work – which, estimated as at least £1.5 billion/US$2 billion per year, every year, is beyond their budgets – lion range states will be urged to collectively request funds from world governments and major international institutions, including UNESCO and the World Bank. He estimates that funding from the international community needs to be secured in the next five to ten years to safeguard the future of Africa's lions.

Of course, protecting lions isn't just the preserve of international conservation organisations and charities. Lion range states are coming together to reverse the trend in declining lion numbers. On 30 May 2016, representatives from African countries that still have lions met in Entebbe, central Uganda. Described by the Convention on International Trade in Endangered Species of Wild Flora and Fauna (CITES) as a historic move, the two-day meeting brought together all of Africa's twenty-eight lion range states[*] – for the first time ever – in common cause to save their lions.

At the meeting, convened jointly by CITES and the Convention on the Conservation of Migratory Species of Wild Animals (CMS), representatives agreed to generate and use action plans, making clear each country's lion conservation strategy and required actions, linked to key dates to make the objectives outlined below possible:

- *Help lions and people live together* by introducing measures that help local people better protect themselves and their livestock from lion attacks, and by providing communities with a fair share of any financial benefits gained by having lions. Inspire local people to join the fight to save lions as they start to benefit from the advantages that lions can bring.
- *Protect the land lions live on* by setting up adaptable and workable ecosystem and wildlife-based land use practices, including measures to thwart agricultural and mining operations that threaten lion habitat.
- *Give lions better legal protection* by bolstering current lion protection legislation and reinforcing law enforcement activities, especially with regard to poaching and the encroachment of protected land by pastoralists.
- *Monitor lion numbers* by improving the collection of scientific data, keeping lion census data accurate and up to date, enabling progress to be measured effectively.

---

[*] Angola, Benin, Botswana, Burkina Faso, Cameroon, the Central African Republic, Chad, Côte d'Ivoire, Ethiopia, Ghana, Guinea, Guinea-Bissau, Kenya, Malawi, Mali, Mozambique, Namibia, Niger, Nigeria, Senegal, Somalia, South Africa, Sudan, Swaziland, Togo, Uganda, Zambia and Zimbabwe.

- *Make lion populations less fragmented* by working with bordering lion range states to restore or create transboundary wildlife corridors.

After the meeting, George Owoyesigire, assistant commissioner for wildlife conservation at the Ugandan Ministry of Tourism, Wildlife and Antiquities, related that the meeting had been 'remarkably constructive' and was for the first time bringing together countries that still had lions, and demonstrated how 'a combination of local and regional efforts as well as regulation of international trade by CITES and enhanced cooperation through CMS can promote the conservation of the species and its habitats'.

Just as with the Cecil Summit's commitments and plans, when and what happens next depends on those involved in the struggle to save *Panthera leo*. But it's important and positive to note that Owoyesigire, as well as remarking on the need for lion range states to take action, also called for support from outside Africa: 'African lion range states call upon CITES, CMS and other partners to support their efforts to conserve and restore this iconic species across the continent.'

In addition to the strategies put in place to help secure a future for Africa's last lions, some scientists have been researching and evaluating another conservation approach that could help revive lion populations – especially those most threatened in West and Central Africa, which aren't part of WildCRU's and Panthera's six priority lion conservation areas.

For almost fifty years, there have been conversations about reintroducing lions to North Africa, where they have been extinct for almost seventy years. Known colloquially as Barbary, Atlas or Moroccan lions, and made distinctive by their thick, shaggy manes and solid, muscle-laden bodies, these charismatic and handsome big cats disappeared from

Algeria, Libya, Morocco and Tunisia as a result of habitat loss and hunting.

Although gone from the wild, there have been informed suggestions that North African lions have not died out completely, because some of their descendants might be found in captivity. Since the late 1990s, several scientists and zoo practitioners have examined the history of captive lions originally from Morocco in North Africa.

Simon Black, a research associate at Durrell Institute of Conservation and Ecology at the University of Kent, and Nobuyuki Yamaguchi, associate professor of animal ecology at the University of Qatar, pulled together information from zoo archives, reports and other literature to create a North African lion stud book, with support from Adrian Harland, animal director at the Port Lympne Reserve in the UK. The book traces the lineage of captive lions believed to be Barbary or Morocco at a zoo in Rabat, the country's capital.

For centuries, the sultans of Morocco kept wild lions as pets. Taken from the Atlas Mountains, these lions were housed in the sultans' beautiful palace gardens in the old capital city of Fez. As lions went extinct in the wild in Morocco and the rest of North Africa, the sultans kept their lions – often referred to as the Moroccan Royal Lion Collection – safe from harm. They remained royal property until 1970, when King Hassan II moved them from his palace to a purpose-built zoo in Rabat.

Since then, Rabat Zoo has shared its lions and cubs with other zoos, mainly in Europe, but also in the United States. Fortunately, the zoo kept records of its lions' movements from the late 1960s to the late 1980s. Using this archive, as well as detailed breeding records from the zoos that received Rabat's lions from 1974 onwards, Black and his colleagues established a stud book for all related lions kept in the captive collection since the 1960s.

This book, called the Moroccan Royal Lion Stud Book, identifies and locates around sixty descendants of King

Hassan's original lions now living in European zoos, plus
twenty-five that remain in Rabat. It also shows that captive
lions currently held in Europe can be traced back to four
lineages from animals originally housed in zoos based in the
Czech Republic, Morocco, Spain and the United States.

The stud book is now being used as part of a captive North
African lion breeding programme, overseen by Adrian
Harland at the Port Lympne Reserve for the European
Association of Zoos and Aquaria (EAZA), who promote
international co-operation between zoos to further wildlife
conservation.

As a result, male and female lions with the Moroccan royal
lineage have been exchanged between a number of zoos,
establishing breeding pairs in Erlebnis Zoo in Germany,
Olomouc Zoo in the Czech Republic and Port Lympne
Reserve in England. To date, cubs have been born in the
German and Czech zoos.

However, before the official captive breeding programme
began, these lions could have been bred with lions from other
regions in Africa. To find out if this had been the case, it was
thought that DNA tests could be carried out to determine if
their ancestors were North Africa's original lions. Unfortunately, this isn't possible because there isn't a viable North
African gene against which to test current samples. All there
is to make a correct identification are DNA scraps extracted
from a handful of bones and teeth from museum samples of
North African lions hundreds of years old.

Not only this, but the most reliable fragment of DNA
sequence that differentiates North African lions from other
lions is tiny. It's a difference so minuscule that Black has
described it as like having to compare one person with
another, including all their similarities and differences, just by
looking at their earlobes.

But the greater challenge is that this fragment of disting-
uishing DNA sequence is only passed down a maternal line
by female lions. So even if there is North African heritage
passing down through males over many generations, it won't

show up in DNA testing. Also, a female whose North African father has passed his genes on to her will not show up.

To complicate matters further, a female lion could test positive as North African even though her father was not. A test of absence of the DNA sequence is not a test of absence of North African DNA, and a test that shows presence of the sequence does not guarantee a 'pure' North African lion!

Until a genetic test is developed that can work with these trickeries, the Moroccan Royal Lion Stud Book and lions in the official captive breeding programme are the best indicators for identifying lions carrying North African genes. Breeding lions like this helps create a genetic safety net. It avoids further inbreeding and also keeps some likely North African lion genes in circulation, without it mixing with the genes of lions from elsewhere in the continent. The descendants of the Moroccan Royal Collection are the closest living link to wild North African lions we have. If nothing is done, the gene could die out and be gone for ever.

So, while the waiting continues for a definitive scientific test to identify the North African lion gene, the animals that are most likely to be direct maternal and paternal descendants from animals held in the long-term collection of the Moroccan royal family continue to be bred. Ensuring the survival of these lions offers the best chance of preserving possible North African bloodlines.

Only once a suitable DNA test is available can we know for definite if North African lion genes still survive in the stud book lions. If future tests show that the gene continues, then the 'ingredients' for a North African lion are still with us. Such lions and their offspring with the 'right' DNA could be 'bred back' – removing any non-North African traits along the way – to eventually produce 'pure' North African lions, which could eventually be returned to their ancestral home.

The possibility of doing this has been mooted on and off since the 1970s. In his recent paper 'The Challenges and Relevance of Exploring the Genetics of North Africa's "Barbary Lion" and the Conservation of Putative Descendants in Captivity' published in the *International Journal of Evolutionary Biology* in July 2016, Black points out why a reintroduction of North African lions needs to be considered:

> Clearly, as apex predator, the species had an important role in North African ecosystems and its cultural importance within North African countries and their near neighbours around the Mediterranean suggests that the lion is a potential flagship for conservation of the Maghreb region. To date, the insufficient evidence to count or discount the relevance of putative descendants, particularly the captive Moroccan Royal Lions, does not change this expectation.

Now that the IUCN recognises that lions from North, Central and West Africa, as well as India, share enough ancestry to form a single evolutionary significant unit (ESU) – known as *Panthera leo leo* – the wider role of captive-bred putative North African big cats in lion conservation across the continent can be considered more seriously. Because the number of wild lions that share the *Panthera leo leo* ESU is so low – just over 520 in India plus the few hundred wild lions hanging precariously onto life in West and Central Africa – any other captive-bred lions with the same DNA could boost the clade's chances of survival.

In his paper, Black notes that the ninety or so lions listed in the EAZA stud book 'represent a population ... perhaps only one-half the size of the wild population in West Africa and one-quarter the size of the wild population in India'. This puts into stark relief the pitifully low numbers of lions left in Africa and India, and strengthens significantly the case for continuing to captive breed North African lions and find a way to prove or disprove their heritage definitively. As Black states: 'If the breeding of Moroccan lions in captivity is deemed worthwhile, by inference and precedent in conservation practice, that breeding programme should be purposed towards the ultimate reintroduction of the animals into the wild. This could ultimately influence the preservation of lions across their precarious northern ranges in India, West Africa, Central Africa, and, potentially, North Africa itself.'

If, in time, representatives of North Africa's lions are found in zoo collections, then their reintroduction to their former North African strongholds in Morocco, Algeria or Tunisia could be seriously considered as a means of preserving *Panthera leo leo* in the wild. Although no official reintroduction project currently exists,* scientists, conservationists and zoological institutions are currently sharing ideas that could lead to one.

This requires ongoing discussions with national and regional government representatives, local NGOs and wildlife conservation bodies to generate interest and get to grips with all that's needed to make a reintroduction project successful and sustainable. It includes ensuring that people living near possible reintroduction sites are on board, and are prepared and able to have large carnivores as neighbours.

Should future genetic testing prove that lions in the North African captive breeding programme aren't part of the northern clade of lions (*Panthera leo leo*), it's still possible that the roar of lions could be heard in Africa's north again.

As lions in India are now classified as *Panthera leo leo*, it could be possible for lions from India to be translocated to repopulate designated areas of North Africa. Such a move would create an additional wild population and further safeguard the species.

If new genetic research shows that lions in the EAZA North African lion captive breeding programme are related to wild lions in India, West and Central Africa, then a new perspective can be taken on their potential use as a genetic pool to support the recovery of the most threatened populations of lions in West and Central Africa.

It could be that these captive lions hold a unique genetic heritage for the global lion population and represent 'a

---

* WildLink International, the most recent project with plans to reintroduce the Barbary lion to Morocco, stopped operating in the 1990s.

considerable proportion of genetic diversity within the species' that may help save all of the world's lions, but most especially those in India and West and Central Africa.

For now, the fate of our lions hangs in the balance. If a DNA test is developed and it definitively proves that we still have North African lions in captivity, in time a population could be reintroduced to North Africa, as well as provide a genetic safety net for vanishing *Panthera leo leo* elsewhere in the continent.

Hopefully, too, global finance – £1.5 billion/US$2 billion a year, every year – will come through, enabling WildCRU and Panthera to restore, manage and provide premium protection for their six priority lionscapes, enabling lion populations to reach their ecological potential, benefitting not just lions but all wildlife and local people, too.

In the meantime, it's a waiting game. But as we wait, lions continue to lose their lives. The war continues. There is no armistice. The human population keeps growing and agriculture continues to expand, further restricting the range of lions. The bushmeat trade shows no sign of abating, and poachers continue to take lions' wild prey, creating more conflict with people when lions take livestock instead.

Until funds start arriving in Africa, lions are in limbo, a conservation twilight zone where they disappear from plain sight on a daily basis. For example, in South Africa, where Africa's most protected lions live, the country's press reported the escape of four male lions from the Kruger National Park in July 2017.

The lions wandered onto a farm in the neighbouring Mara Valley and killed a cow. In response, one lion was shot and killed and another injured by the cow's owner. Unfortunately, because of difficult circumstances, wildlife rangers were unable to dart and retrieve the surviving lions and were forced to destroy them in the interest of public safety. Four male lions gone just like that in what is probably the continent's most lion-friendly country.

As well as being a deadly wait, it's an anxious one, too. Even if there is a global pledge to provide the funding needed to fulfil WildCRU's call to arms to 'secure the six and save the sixty', promises, despite best intentions, aren't always honoured. For example, in 1992, developed nations promised £1.5 billion/US$2 billion a year, every year – the amount that lions need – to help developing nations with their conservation efforts. That money has never come through.

However, funding has been secured for Panthera and WildCRU to host another Cecil Summit. This will take place in Africa in 2018. Delegates are expected to include representatives from all the continent's lion range states as well as a potent mix of experts and innovative thinkers from all over the world. The summit will follow the same 'Oxford format' as the 2016 event.

CHAPTER EIGHT

Beyond Gold

And the desert wild
Become a garden mild

William Blake, 'The Little Girl Lost',
Songs of Innocence and of Experience

In 2016, a year after the death of Cecil, Panthera president and chief conservation officer Luke Hunter said: 'With the loss of Cecil, the world responded unequivocally that it stands with Africa in saving the lion. Sadly, we have since lost hundreds and possibly thousands of lions. The species is now approaching the point of no return in many countries.'

The countries that Hunter referred to are in West Africa – in Benin, Burkina Faso, Niger, Nigeria and Senegal – where human population growth has simply left no room for lions. Between them these countries are home to just 400 or so of the region's last remaining lions.

According to the United Nation's 2017 world population figures, all five countries, especially Nigeria, are experiencing rapid population growth. For example, Nigeria, which has around 340 lions, is one of the fastest-growing populations in the world. In 2017, the population was estimated at 191 million. By 2050, it's predicted to rise to 411 million, which, the United Nations says, will make it the third most populous country on the planet.

Collectively, the populations of these West African lion range states are expected to increase from nearly 260 million in 2017 to over 580 million in 2050, reaching more than a billion by 2100. Four of the five countries – Benin, Burkina

Faso, Niger and Senegal – are among the world's least developed countries. It's the poorest places, the United Nations notes, where population growth is most concentrated, making it more difficult for governments 'to eradicate poverty, reduce inequality, combat hunger and malnutrition, expand and update education and health systems, improve the provision of basic services and ensure that no one is left behind'.

These are countries unable to provide for their people, let alone their wildlife. For populations struggling to survive, where the region's economy is depressed and fish stocks depleted, bushmeat is keeping people alive.

Consequently, poaching is rife. Speaking with *National Geographic* in 2014, Philipp Henschel, a Panthera lion programme co-ordinator based in West Africa, said the activity was now so orchestrated that when working in Burkina Faso, his team saw 'poachers coming from Nigeria, 100 miles away, to shoot big animals and carry them across the border in pickup trucks'.

Henschel also noted the chronic underfunding of protected areas in the region, adding that of the twenty-one management areas his team looked at, six of them had no 'operating budget at all' and all were understaffed. Effectively they were paper parks 'systematically being stripped by poachers'.

Lions here don't stand a chance. Their prey has been wiped out by poachers. Their habitats have been lost to farms and settlements. And when they kill livestock, they come into deadly conflict with people.

These fragile lion populations, known among conservationists as 'the living dead', are imperilled – too tiny and too remote to survive much longer. Their predicament is so dire that they are categorised as critically endangered by the IUCN. For them, extinction seems inevitable. In West Africa, it seems the war on lions has been won.

No areas in West or Central Africa are included on WildCRU's and Panthera's six prioritised lionscapes as part of their 'secure the six and save the sixty' initiative. Of course they're not. Conservationists have to prioritise. For lions in

West Africa and many in Central Africa, the crisis has gone too far – there is little point in throwing a life raft to a species already drowned.

For a lionscape to be included in the priority six, it needs to be a well-protected area where lions are already paying their way. It needs to be a place where there are enough lions and other wildlife to attract tourists, whose money in turn provides the means to pay for continuing protection.

In January 2017, Paul Funston, a senior lion programme director with Panthera, commented that such areas are few and far between, but where they do exist, their conservation value should not be underestimated. Funston described them as 'pure gold', places 'where tourists can see lions really being lions in all the amazing facets of their behavior, and where lions properly fulfil their ecological role'.

These golden places just don't exist in West and Central Africa. In these regions, national parks and reserves can be difficult and expensive to reach. They are often uncomfortable to be in and perceived as potentially dangerous places. They are off the tourist map. People don't want to go on holiday there, especially if seeing a lion or any other Big Five animal is unlikely. Tourists don't visit because there's not enough infrastructure or wildlife, and operators won't invest because no one wants to go there. It's a conservation catch-22.

Six of the seven countries that are home to the priority six areas are in southern Africa – Botswana, Mozambique, Namibia, South Africa, Zambia and Zimbabwe.* Here, lion numbers are increasing. Together, Botswana, Namibia, South Africa and Zimbabwe are home to an estimated 24 to 33 per cent of Africa's lions, where populations have increased by 12 per cent in the last twenty or so years. In South Africa, they are doing so well that the IUCN has declared the country's lions to be of least concern regarding extinction.

---

* The seventh country is Tanzania in East Africa.

These countries are generally well known to tourists. They are relatively easy to reach and offer out-of-the-ordinary experiences, alongside all the recognisable mod cons and comforts. National parks and private reserves are carefully managed for long-term viability, ensuring the pleasure and comfort of their safari guests and making them hotspots for a wide range of tourists, not just wildlife enthusiasts.

Luxurious accommodation is widely available. Champagne sundowners and candlelit dinners under the stars, all *Out of Africa* perfect, are easy to book. Guided safaris in comfortable jeeps take visitors safely into the bush where the habituation and abundance of wildlife means that seeing lions and other Big Five animals is almost guaranteed.

In a piece for the *New York Times* published in 2015, nature writer and author Helen Macdonald shares her experience of a safari at the Kruger National Park in northeast South Africa. Out on a game drive, Macdonald's jeep stops close by a male lion that has just woken up. Observing the lion, Macdonald says she is unnerved, and that although she knows the lion won't attack, she wonders if the group should 'acknowledge that he might?'. Macdonald is so close to the lion, she can see (without binoculars) a wound on his face, moving on to say that she feels 'vaguely betrayed by his proximity' and thinks that she wants 'it to be harder to see a lion'.

For Macdonald, her first encounter with a wild lion feels slightly disappointing. Getting so close to a lion so quickly and with such little effort feels like cheating. Shouldn't such intimacy be the reward for years of slow, steady contact? Macdonald also finds the experience a little unreal, as if she is watching wildlife documentary footage, especially when her guide's 'expert tone' is reminiscent of 'voice-over narration'.

The lion appears unfazed by the people in the jeep photographing him, paparazzi-style. He even has a name: Terminator. Of course, Terminator, like most of the other wildlife in Kruger, is highly habituated. 'All of the animals here have already been born with the vehicles,' Macdonald's guide explains.

Many protected areas in southern Africa, especially South Africa, are incredibly well managed by professionals dedicated to ensuring the survival of the animals they steward. Much of the funding that pays for this comes from tourists who want to see animals in the wild, just as they've seen them on TV.

In the journal *Geoforum*, British geographer Andrew Norton describes how, in his view, animals in national parks, with their habitats and populations carefully managed, have become domesticated 'through exposure to tourists rather than caging; and the experience is a constructed one which offers access to a "wild", primeval nature'.

There is a sense, Norton maintains, that animals seen on safari are performing, providing a show for audiences in jeeps. In this vein, it's interesting to note that the Madikwe Game Reserve in South Africa that I visited describes itself as a 'wildlife theatre' on its website.

Back at Kruger, Macdonald's guide tells her how visitors love to watch the park's predators because not only are they beautiful, 'they are entertaining and interesting' and 'they'll stalk, they'll fight, they'll mate'. Because so many protected large areas in southern Africa are fenced in, some conservationists refer to them as 'megazoos'.

In her book *Wild Life: The Institution of Nature*, Irus Braverman, professor of law and adjunct professor of geography at the University at Buffalo, the State University of New York, explores the narrowing gap between conserving species *ex situ* and *in situ*. As part of this process, Braverman cites conversations she's had with professionals involved in both kinds of conservation projects around the world.

For example, Hamish Currie, director of Back to Africa, a South African organisation reintroducing species to national parks around Africa, tells Braverman that in his view 'there are very few places left that are really "wild" '.

Africa is too often romanticised, he says, and that when people 'think about this vast continent of Africa' they visualise huge open spaces where animals travel 'vast distances' and where 'genetic exchange is taking place', but the reality is that the wild is actually 'being boiled down to smaller and smaller pockets where animals might have to be managed'.

In southern Africa, many of these wild pockets are enclosed by fences. Within them, the captive wild is kept running smoothly by ensuring that populations of carnivores and herbivores, as well as plant species, are carefully balanced to maintain the desired status quo, preserving a contained and fully functioning ecosystem.

It's this management of the captive wild – from the micro-management of habitats and control of the fecundity of fauna, to the carefully engineered installation of fencing – that many believe is saving countless lions' lives in the south, in South Africa particularly, and helping to reinvigorate the region's lion populations.

Without fences to protect them, many conservationists maintain, lions are just too vulnerable to survive. Fences stop lions coming into conflict with local people. They keep poachers out, safeguard lion prey and protect lion habitat from encroachment.

'In South Africa,' Peter Dollinger, director emeritus of the World Association of Zoos and Aquariums, tells Irus Braverman in *Wild Life*, 'there is no single wild lion.' And he's probably right. Lions in South Africa are national assets. They're investments, bringing in income and paying their own way. Like gold in bank vaults, they're guarded and protected in thriving – if not evolving – sealed ecosystems.

Much of southern Africa protects its lions. They're worth it. People pay good money to see them. They're wildlife gold, the very opposite of their smaller cousins going terminal in Central and West Africa. In the countries that tourism forgot, governments simply can't afford to properly protect their

national parks. Fences are prohibitively expensive and are few
and far between.

Every day a little bit more of the wild in Africa's west shrinks
away. As it does, the life that depends on it goes too.
Disappearing lions are indicators of whole ecosystems under
threat. Harbingers of extinction, they are the grim roaring
reapers that tell of a crumbling, dismantled biodiversity, the
loss of life and the fast-vanishing wild.

When large mammals like lions reach the end of the line,
countless other species join them – they've lost their food
sources and habitats too. All life is affected by the waning
wild; insects, amphibians, birds and mammals slip away too.
International funding, if it does come, will come too late to
save much of the wildlife here.

In *A Field Guide to Getting Lost*, a collection of essays about
loss published in 2006, writer Rebecca Solnit talks about the
fragility of the existence of animals and what their
disappearance from the world means to her. She begins by
quoting American philosopher Henry David Thoreau: 'They
are all beasts of burden, in a sense, made to carry some
portion of our thoughts,' and then adds her own thoughts:
'Animals are the old language of the imagination; one of the
ten thousand tragedies of their disappearance would be a
silencing of this speech.'

Central and West Africa are becoming quieter places,
muted as wildlife is snuffed out. The roar of lions in many of
these countries won't be heard again. These are the lions
roaring their final roars, a soundtrack from a wilder past,
never to be heard again. They are the lions that the tourists
never went to see, that nobody photographed. Forgotten and
soon to be gone, almost as if they had never existed.

The only lion-filled wild we have left is the captive wild, a
pocket in Tanzania in the east, the rest down in the south.
But maybe that's acceptable. It's safer and easier for us to visit

the captive wild, to see lions roam land carefully restored to match the Africa of our safari-fuelled imaginations. Here, the wild is wild enough. There's just the right amount of blood, tooth and claw to make our hearts beat that little bit faster than they do when we visit our own worn-out, overgrazed national parks.

The captive wild is still beautiful, it still looks like and smells like Africa. Parks and reserves are so huge, fences are rarely seen. Spotting a lion is no less a thrill because it's habituated and inside a fenced-in national park. The lions are still handsome, the males magnificently maned, unlike their cousins in the west. They are the lions of our imaginations, the storybook lions and the Trafalgar Square lions we want big cats to look like. And we can get close, really close, to these creatures. We can capture their image on our cameras, digital trophies we can take home and share with family and friends, as well as strangers, all over the world.

In the closing paragraph of her piece for the *New York Times*, Helen Macdonald says that although her initial feelings about her lion encounter were disappointing, when Terminator roars and lets out 'a vast, low and terrifying sound', she suddenly feels different. As her 'human musings fall away' she finds she is taken 'into the lion's world', where she becomes 'a being without thoughts, a being of flesh and fear, terror and simple awe'.

A species that once roamed freely is now mainly found behind fences in Africa's captive wild. Natural commodities that can pay their way, many lions are protected and managed from cradle to grave. Some find their fertility is precision-engineered; eggs and sperm monitored, ovulation and ejaculation perfectly planned and timed. Today they are the stars of conservation stories, rather than magic and fable.

The lions that remain – living souvenirs from a wild past – are kept safe in managed pockets of wildness. Maintaining their homes is a forever commitment; to protect lions from us, we must intervene in perpetuity.

Even if international funding doesn't come through to 'secure the six to save the sixty', lions in South Africa and in other well-protected and much-visited parks in the south of the continent and Tanzania will survive and continue to grow in number. The same applies for the 500 or so lions at Gir in south-west India. Elsewhere, though, lions will continue to disappear, to die out.

For some, this might be enough lions. People can rest easy knowing that lions won't go completely extinct. There will be significantly fewer than we have now and much of Africa will have none at all, but the species will go on in the captive wild at least. We will still be able to go on holiday and marvel at them, be able to tell our grandchildren that such wonderful animals are still among us.

I imagine these last lions like shadows kept in a beautiful jar. We can't see them until we think about them, but we know they're still there, safe and protected from harm. When we do think of them, when we illuminate them in our imaginations, they're still magnificent, wearing their golden crowns nobly and roaring furiously, but they're no longer the king of the beasts, simply because there is no real wild left for them to truly reside over.

Acknowledgements

Thank you to all of my family, but most especially my husband Conor Jameson, my son Jacob Pake (thank you so much for building me a wonderful website), my mother Deirdre Hume and dear sissie Carrie Betts for your ongoing support and encouragement while writing the book. Not forgetting my lion-hearted friend Isabel King for doing the same.

I would also like to express my gratitude to a top daughter-and-father team comprising Isabel and Daniel King – whom I consider grammar royalty – for sharing their editorial skills as I put the book together. At Bloomsbury, I would like to thank Lisa Thomas for her guidance regarding the book's early chapters, my editor Jenny Campbell for her kindness, support and pertinent advice and Myriam Birch for her copy-editing prowess and expertise.

Likewise, my gratitude goes as well to Tim Jepson for commissioning my first travel piece and kick-starting my travel-writing career and enabling me to encounter wildlife all over the world.

Ross Barnett and Simon Black: I thank you both for your advice and for sharing your extraordinary knowledge about lion evolution and genetics with me. In a similar vein, I would also like to thank David Macdonald and his team at WildCRU for sharing their time, important research findings and enormous conservation expertise. And, of course, gratitude to and admiration for all the hands-on people working in Africa and India to help keep lions alive in difficult and often dangerous circumstances. Lions still roar because of you.

At the Akagera National Park in Rwanda, I send my thanks to Sarah Hall and Jes Gruner, and also Maddy Uwase and Nathan Mwesige, who helped me look for the country's only lions. At the Zoological Society of London (ZSL), my thanks go to Donna Campanella and Gitanjali Bhattacharya, and at the Aspinall Foundation I thank Amanda McCabe. I would also like to thank Salisha Chandra, Paul Thomson and Shavini Bhalla. I'm so sorry I wasn't able to visit your projects in Kenya.

For kind permission to use their wonderful images, I must thank Akagera National Park, the Aspinall Foundation, Patrick Aventurier, the Caverne du Pont d'Arc, Sarah Hall, Dave Rolfe, Daniel Sprawson, David Toovey, Gaël Vande Weghe and ZSL.

I would also like to thank Panthera and Ross Barnett for allowing the cartographer, Brian Southern, to recreate their maps for use in this book.

And, finally, a very special and a very big thank you to the kind and generous people who helped me travel to Rwanda and visit the lion reintroduction project at Akagera National Park. They include: Rebecca Bearn, Carrie Betts, Kevin Betts, Esther Cameron, Phill Capstick, Camila Cavalcante, Kate Chase, Ian Foulsham, Rachel Gill, Dilys Gladwell, Paul Golding, Deirdre Hume, Anita Kerai, Alan McCredie, Patrick Minne, Mike Perry and Mads Petersen. I am also grateful for the support of the Marriott Hotel in Kigali, who demonstrated their support for wildlife by providing a complimentary overnight stay. Likewise, I also thank the Akagera Game Lodge and Tented Camp for their hospitality and discounted rates.

Further Reading

A full list of resources can be found at whenthelastlionroars.co.uk.

Select Bibliography

Attenborough, David. 1987. *The First Eden: The Mediterranean World and Man*. William Collins & Son Ltd, USA

Barnett, R. et al. 2006. Lost Populations and Preserving Genetic Diversity in the Lion *Panthera leo*: Implications for its *Ex Situ* Conservation. *Conservation Genetics* 7:14 507–514

Barnett, R. & Nijman, V. 2007. Using Ancient DNA Techniques to Identify the Origin of Unprovenanced Museum Specimens, as Illustrated by the Identification of a 19th Century Lion from Amsterdam. *Contributions to Zoology* 76, 87–94

Barnett, R. et al. 2014. Revealing the Maternal Demographic History of *Panthera leo* Using Ancient DNA and a Spatially Explicit Genealogical Analysis. *BMC Evolutionary Biology* 14 1–11

Bauer, B. et al. 2015. Lion (*Panthera leo*) Populations are Declining Rapidly across Africa, Except in Intensively Managed Areas. *Proceedings of the National Academy of Sciences* 112:48 14894–14899

Bertola, L. D. et al. 2011. Genetic Diversity, Evolutionary History and Implications for Conservation of the Lion (*Panthera leo*) in West and Central Africa. *Journal of Biogeography* 38:7 1356–1367

Bisset, John Jarvis. 1875. *Sport and War; or, Recollections of Fighting and Hunting in South Africa*. J. Murray, London

Black, S. A. et al. 2013. Examining the Extinction of the Barbary Lion and its Implications for Felid Conservation. *PLOS ONE*

Black, S. et al. 2016. Maintaining the Genetic Health of Putative Barbary Lions in Captivity: An Analysis of Moroccan Royal Lions. *European Journal of Wildlife Research* 56:1 21–31

Blandford, W. T. 1888–1891. *The Fauna of British India, Including Ceylon and Burma: Mammalia.* Taylor and Francis, London

Blasco, R. et al. 2010. The Hunted Hunter: The Capture of a Lion (*Panthera leo fossilis*) at the Gran Dolina Site, Sierra de Atapuerca, Spain. *Journal of Archaeological Science* 37:8 2051–2060

Bomgardner, David, L. 2000. *The Story of the Roman Amphitheatre.* Routledge, Oxon

Braverman, Irus. 2014. *Wild Life: The Institution of Nature.* Stanford University Press, California

Bull, Bartle. 1988. *Safari: A Chronicle of Adventure.* Viking, London

Burchell, William John. 1822. *Travels in the Interior of Southern Africa.* Longman, Hurst, Rees, Orme & Brown, London

Cattrick, Alan. 1959. *In Spoor of Blood.* Howard Timmins, Cape Town

Cueto, M. et al. 2016. Under the Skin of a Lion: Unique Evidence of Upper Paleolithic Exploitation and Use of Cave Lion (*Panthera spelaea*) from the Lower Gallery of La Garma (Spain). *PLOS ONE*

Day, David. 1983. *The Doomsday Book of Animals: An Illustrated Account of the Fascinating Creatures which the World Will Never See Again.* The Viking Press, New York

Diedrich, C. G. 2011. Late Pleistocene *Panthera leo spelaea* (Goldfuss, 1810) Skeletons from the Czech Republic (Central Europe); their Pathological Cranial Features and Injuries Resulting from Intraspecific Fights, Conflicts with Hyenas, and Attacks on Cave Bears. *Bulletin of Geosciences* 86:4 817–840

Diedrich, C. G. & Copeland, J. P. 2010. Upper Pleistocene Gulo Gulo (Linné, 1758): Remains from the Srbsko Chlum-Komin Hyena Den Cave in the Bohemian Karst, Czech Republic, with Comparisons to Contemporary Wolverines. *Journal of Cave and Karst Studies* 72: 2 122–127

Divyabhanusinh. 2005. *The Story of Asia's Lions.* Marg Publications, India

Divyabhanusinh. 2006. Junagadh State and its Lions: Conservation in Princely India, 1879–1947. *Conservation and Society.* 4 522–540

Divyabhanusinh. 2006. The Great Mughals Go Hunting Lions. *Environmental Issues in India: A Reader*. Pearson, Kindle

Elliot, N. B. & Gopalaswamy, A. M. 2017. Toward Accurate and Precise Estimates of Lion Density. *Conservation Biology* 31:4 934–943

Funston, P. et al. 2017. *Beyond Cecil: Africa's Lions in Crisis.* Panthera & WildAid

Gérard, Jules. 1856. *The Life and Adventures of Jules Gérard the Lion-Killer.* Lambert and Co., London

Gordon-Cumming, Roualeyn. 1874. *Five Years of a Hunter's Life in the Far Interior of South Africa.* Harper & Brothers, New York

Groom, R. J., Funston, P. J. & Mandisodza, R. 2014. Surveys of Lions *Panthera leo* in Protected Areas in Zimbabwe Yield Disturbing Results: What is Driving the Population Collapse? 48:3 385–393

Hayward, M. & Somers, M. 2014. Translocations in South Africa: Lion Reintroduction in Perspective. *Current Conservation* 8:4 12–19

Hazzah, L. et al. 2014. Efficacy of Two Lion Conservation Programs in Maasailand, Kenya. *Conservation Biology*. 28:5 851–860

Henschel, P. et al. 2010. Lions Status Updates from Five Range Countries in West and Central Africa. *Cat News* 52

Henschel, P. et al. 2014. The Lion in West Africa is Critically Endangered. *PLOS ONE*

Hunter, J. A. 1986. *Hunter: The Adventurous Life of One of the Greatest of Africa's Hunters.* Tideline Books, Wiltshire

Jackson, Deirdre. 2010. *Lion.* Reaktion Books Ltd, London

Jennison, George. 2005. *Animals for Show and Pleasure in Ancient Rome.* University of Pennsylvania Press, Philadelphia

Kirillova, V. et al. 2015. On the Discovery of a Cave Lion from the Malyi Anyui River (Chukotka, Russia). *Quaternary Science Reviews* 117 135–151

Kitchener, A. C. et al. 2017. A Revised Taxonomy of the Felidae. The Final Report of the Cat Classification Task Force of the IUCN/SSC Cat Specialist Group. *Cat News* Special 11 1–80

Kolbert, Elizabeth. 2014. *The Sixth Extinction: An Unnatural History*. Picador, New York

Layard, H. 1887. *Early Adventures in Persia, Syria and Babylonia*. John Murray, London

Lindsey, P. A., Roulet, P. A. & Romanach, S. S. 2007. Economic and Conservation Significance of the Trophy Hunting Industry in Sub-Saharan Africa. *Biological Conservation*. 134 455–469

Lindsey, P. A. 2008. Trophy Hunting in Sub Saharan Africa: Economic Scale and Conservation Significance. *Best Practices in Sustainable Hunting*. International Council for Game and Wildlife Conservation. Hungary

Lindsey, P. A. et al. 2012. The Significance of African Lions for the Financial Viability of Trophy Hunting and the Maintenance of Wild Land. *PLOS ONE*

Lindsey, P. et al. 2015. *Illegal Hunting and the Bush-meat Trade in Savanna Africa: Drivers, Impacts and Solutions to Address the Problem*. Report for the Food and Agriculture Organization of the United Nations. New York

Lindsey, P. A. et al. 2017. The Performance of African Protected Areas for Lions and their Prey. *Biological Conservation* 209 137–149

Loftus, W. 1857. *Travels and Researches in Chaldæa and Susiana; with an Account of Excavations at Warka, the 'Erech' of Nimrod, and Shúsh, 'Shushan the palace' of Esther, in 1849–52*. Robert Carter & Brothers, New York

Macdonald, David, W. 2016. *Report on Lion Conservation with Particular Respect to the Issue of Trophy Hunting*. University of Oxford

Macdonald, David, W. and Loveridge, A. 2010. *The Biology and Conservation of Wild Felids*. Oxford University Press

Mauricio, Anton. & Salesa, M. J. 2012. Walking with Lions: The Missed Potential of Quaternary Palaeoichnology. *Quaternary Science Reviews* 49 106–108

McClung, Robert M. 1976. *Lost Wild Worlds*. William Morrow and Company, New York

Miller, S. et al. 2016. A Conservation Assessment of *Panthera leo*. Included in The Red List of Mammals of South Africa,

Swaziland and Lesotho. South African National Biodiversity Institute and Endangered Wildlife Trust, South Africa

Mitra, Sudipta. 2005. *Gir Forest and the Saga of the Asiatic Lion*. Indus Publishing Company, New Delhi

Norton, A. 1996. Experiencing Nature: The Reproduction of Environmental Discourse Through Safari Tourism in East Africa. *Geoforum* 27:3 355–373

Omoya, E. O. et al. 2013. Estimating Population Sizes Lions *Panthera leo* and Spotted Hyenas *Crocuta crocuta* in Uganda's Savannah Parks, Using Lure Count Methods. *Oryx* 48:3 394–401

Packer, Craig. 1992. Captives in the Wild. *National Geographic*. April 122–136

Packer, Craig. 2015. *Lions in the Balance: Man-Eaters, Manes and Men with Guns*. University of Chicago Press, Chicago

Packer, C. et al. 2010. The Effects of Trophy Hunting on Lion and Leopard Populations in Tanzania. *Conservation Biology* 25:1 142–153

Pease, Alfred. 1914. *The Book of the Lion*. John Murray, London

Quammen, David. 2005. *Monster of God: The Man-Eating Predator in the Jungles of History and the Mind*. Pimlico, London

Rawlinson, George. 1897. *The Story of Ancient Egypt*. Unwin, London

Riggio, J. et al. 2012. The Size of Savannah Africa: A Lion's (*Panthera leo*) View. *Biodiversity and Conservation* 22:1 17–35

Singh, H. S. & Gibson, L. 2011. A Conservation Success Story in the Otherwise Dire Megafauna Extinction Crisis: The Asiatic Lion (*Panthera leo persica*) of Gir Forest. *Biological Conservation* 144:5 1753–1757

Solnit, Rebecca. 2006. *A Field Guide to Getting Lost*. Canongate, Edinburgh

Stuart, A. J. & Lister, A. M. 2011. Extinction Chronology of the Cave Lion *Panthera Spelaea*. *Quaternary Science Reviews* 30:17–18 2329–2340

Venkataraman, M. 2014. Conservation of Asiatic Lions: Where Do We Go Now? *Current Conservation* 8:4 24–29

Venkataraman, M., Macdonald, D. W. & Montgomery, A. 2014. Managing Success: Asiatic Lion Conservation, Interface

Problems and People's Perceptions in the Gir Protected Area. *Biological Conservation* 174:6 120–126

Williams,V. L., Loveridge, A. J. & Macdonald, D.W. 2015. *Bones of Contention: An Assessment of the South African Trade in African Lion* Panthera leo *Bones and Other Body Parts.* TRAFFIC, Cambridge, UK & WildCRU, Oxford

Wilson, A. 1941. *South West Persia: A Political Officer's Diary 1907–1914.* Oxford University Press, London

Yamaguchi, N. & Haddane, B. 2002. The North African Barbary Lion and the Atlas Lion Project. *International Zoo News* 49:8 465–482

Websites

Below is a list of websites belonging to projects and organisations committed to lion conservation.

African Lion & Environmental Research Trust: lionalert.org
African People & Wildlife: africanpeoplewildlife.org
African Wildlife Foundation: awf.org/wildlife-conservation/lion
Big Cats Initiative: nationalgeographic.org/projects/big-cats-initiative
Born Free Foundation: bornfree.org.uk
Captured in Africa Foundation: capturedinafricafoundation.com
Elsa Conservation Trust: elsamere.com/elsa/conservation-trust
Ewaso Lions: ewasolions.org
Let Lions Live: letlionslive.org
LionAid: lionaid.org
Lion Conservation Fund: lionconservationfund.org/center.html
Lion Guardians: lionguardians.org
Lion Research Center: cbs.umn.edu/research/labs/lionresearch
Living with Lions: livingwithlions.org
KopeLion: kopelion.org
Ruaha Carnivore Project: ruahacarnivoreproject.com
Panthera: panthera.org
Predator Conservation Fund: lionconservationfund.org/center.html

Protecting African Lions: protectingafricanlions.org
WildCRU: wildcru.org
Wildlife Conservation Trust: asiaticlion.org
Zoological Society of London's Asiatic lions Campaign: zsl.org/
    support-us/zsls-asiatic-lions-campaign

Index

Index